LCD Backlights

Wiley-SID Series in Display Technology

Series Editor:
Anthony C. Lowe

Consultant Editor:
Michael A. Kriss

LCD Backlights

Edited by

Shunsuke Kobayashi

Tokyo University of Science, Yamaguchi, Japan

Shigeo Mikoshiba

The University of Electro-Communications, Tokyo, Japan

Sungkyoo Lim

Dankook University, Korea

A John Wiley & Sons, Ltd., Publication

This edition first published 2009
© 2009, John Wiley & Sons, Ltd

Registered office
John Wiley & Sons Ltd, The Atrium, Southern Gate, Chichester, West Sussex, PO19 8SQ,
United Kingdom

For details of our global editorial offices, for customer services and for information about
how to apply for permission to reuse the copyright material in this book please see our
website at www.wiley.com.

Library of Congress Cataloguing-in-Publication Data

LCD backlights / edited by Shunsuke Kobayashi, Shigeo Mikoshiba, Sungkyoo Lim.
    p.   cm.
  Includes bibliographical references and index.
  ISBN 978-0-470-69967-6 (cloth)
  1. Liquid crystal displays–Equipment and supplies.   2. Electric lamps.   I. Kobayashi,
Shunsuke.   II. Mikoshiba, Shigeo.   III. Lim, Sungkyoo.
   TK7872.L56.L33 2009
   621.3815′422–dc22

                                                                              2009004103

A catalogue record for this book is available from the British Library.

ISBN: 978-0-470-69967-6

Set in 10 on 12.5 pt Palatino by SNP Best-set Typesetter Ltd., Hong Kong
Printed in Great Britain by TJ International Ltd, Padstow, Cornwall

Contents

PART THREE   OPTICAL COMPONENTS

Series Editor's Foreword

Liquid Crystal Displays dominate the markets for mobile and monitor displays and as their market share grows inexorably in TV and the Digital Out Of Home (DOOH) signage markets, there has probably never been a more appropriate moment than this to add a book on LCD Backlights as the seventeenth to be published in this Wiley-SID Series.

The range of technologies and applications to which they are applied which is covered by this book is very broad, so it is appropriate that this should be an edited volume with contributions from leading experts from Asia and Europe. Of course, in keeping with the great mass of manufacturing being centred in Asia, most of the contributions are from that region.

This book was originally published in Japanese and in preparing the current edition, some limited editing has been carried out to bring certain time-variant information up to date. It has been edited by three well-known and highly respected people in the field, Professors Kobayashi, Mikoshiba and Lim. They have gathered input from twenty experts from eighteen companies and institutions which they have organised into three sections covering Backlights by Use, Light Source Devices and Optical Components, a manner of organisation which I believe the reader will find useful.

TV presents possibly the greatest challenge for backlights, requiring the widest colour gamut, the highest contrast ratio and the fastest, blur-free response. Some monitor applications also require wide colour gamut and very high contrast. The first section begins with a discussion of trends and requirements for TV backlights. The issues of TV backlight control, colour range and power are addressed in detail in chapters covering backlights which can be modulated in intensity and hue and which contain light sources which emit at more than three (up to six) separate wavelengths. The section concludes with chapters on notebook PC/monitor and handheld display backlights.

The section on light source devices covers hot, cold, external cathode, flat and mercury-free fluorescent lamps, electrodeless lamps, LEDs, OLEDs, inorganic EL and field emission sources as the light sources for backlights together with descriptions of their means of driving. This is a comprehensive list. Not all the technologies described will become commercially successful, but increasing interest in and legislation about power consumption and the use of toxic or unrecyclable materials may provide the driving force to develop mercury-free and longer lived alternatives to today's market leaders, CCFLs and LEDs.

The final section describes the optical components, light guide plates, diffusers, grating/lens and reflective polarising films, which are required to guide light from the light source to the display with the required uniformity of intensity, polarisation state and angular extent to meet the differing requirements of different applications.

This is a comprehensive and timely book. Of course, the field it addresses is one of rapid development and changing requirements, so, as is true of any book, the reader will need to keep abreast by reading the latest published research papers. This book will provide the reader with a rich and comprehensive grounding in the subject. Its numerous references will provide guidance for anyone who wishes to delve in greater depth into any of the subject areas covered.

I welcome this latest addition to the series. It will prove to be a very useful source of information to all who are concerned with the design, development, production and use of LCDs in all their forms.

Anthony Lowe
Series Editor
Braishfield, UK

About the Editors

Professor Kobayashi is with the Liquid Crystal Institute, Tokyo University of Science, Yamaguchi, Sanyo-Onoda, Yamaguchi 756-0884, Japan.

Professor Emeritus Mikoshiba is with the Department of Electronic Engineering, The University of Electro-Communications, 1-5-1, Chofugaoka, Chofu, Tokyo 182-8585, Japan.

Professor Lim is with the Department of Electronics Engineering, Dankook University, 29 Anseo-dong, Cheonan, Choongnam 330-714, Korea.

List of Contributors

S. Aoyama, *Omron Corporation*

F. Hanzawa, *Sumitomo 3M*

B. H. Hong, *Kwangwoon University*

J. Huttner, *OSRAM opto Semiconductors GmbH*

Y. Ishiwatari, *Asahi Kasei Chemicals*

J. Jang, *Kyung Hee University*

G. S. Kim, *GLD Co., Ltd and Mirae Corporation*

J. H. Ko, *Hallym University*

Y. Kondo, *NEC LCD Technologies*

S. Y. Lee, *Samsung Electronics*

S. Lim, *Dankook University*

F. Okamoto, *Matsushita Electric Works*

S. Okamoto, *NHK Science and Technical Research Laboratories*

T. Shiga, *The University of Electro-Communications*

S. Sluyterman, *Philips Lighting*

H. Sugiura, *Mitsubishi Electric Corporation*

T. Uematsu, *TDK*

M. Ushirozawa, *NHK Science and Technical Research Laboratories*

K. Yamaguchi, *Panasonic Photo and Lighting*

T. Yamamoto, *Hitachi, Ltd*

M. Zeiler, *OSRAM Opto Semiconductors GmbH*

Preface

Liquid crystal displays (LCDs) are used for a wide variety of information displays, for example LC-TVs, personal computers, mobile phones and car navigation systems. Thus, LCDs as electronic display devices have become a part of everyday life. There are two categories of electronic information displays: one is an emissive type such as CRTs, PDPs, ELDs, OLEDs, LEDs and FEDs, and the other is a non-emissive type such as LCDs, electronic papers and hard copy printings. In the case of LCDs there are the direct-view types and the projection types. Further, the direct-view types can be divided into three categories: with backlights, without backlights and a hybrid trans-flective type which operates with or without backlights depending on the environmental lighting conditions.

The flat panel display global market in 2008 was approaching US$ 100 billion. Among this LCDs occupied a 93% share. Most LCD devices need backlight units. There are three important issues for the backlight units: first, a reduction in the backlight power is urgently needed since the power consumption of a backlight unit uses 70 to 80% of the LCD module; second is a reduction in the cost; and finally, to the establishment and maintenance of adequate supplies.

The purpose of this book is to promote research and development, so that better backlight units can be manufactured through improvements to their optical and electronic characteristics. Worldwide authorities have discussed the state-of-the-art technologies for the backlights and their systems in this book. We shall be able to contribute to the 'Save Our Earth' program through 3R (recycle, reduce and reuse) activities. The editors believe that this book, as one of the Wiley–SID book series, will be useful to readers who are interested in flat panel displays. The original Japanese version of the book was published by Science & Technology Co., Ltd, in the Japanese language.

Shunsuke Kobayashi
Shigeo Mikoshiba
Sungkyoo Lim

Part One
Backlights by Use

1

Technical Trends and Requirements/ Specifications for LCD TV Backlights

S. Y. Lee

SAMSUNG Electronics

1.1 Introduction

Information display devices play a major role in the transition to information-oriented and ubiquitous societies. The global move toward information societies is causing a sharp increase in the demand for information display devices. Liquid crystal display (LCD) is competing fiercely with plasma display and digital light processing technologies for the large-screen TV market. Until now, the TV market has been expected to be divided into two segments: plasma technology will dominate large screen sizes – above 30 inches – while LCDs will primarily address the smaller size market. Contrary to this prediction, however, the rapid development of LCD technology and the resulting price competitiveness have allowed it to penetrate into the 40- to 50-inch segment as well as the 30-inch range.[1] Some market researchers

LCD Backlights   Edited by Shunsuke Kobayashi, Shigeo Mikoshiba and Sungkyoo Lim
© 2009 John Wiley & Sons, Ltd.

forecast that the demand for LCDs will outnumber that for PDPs in the 40-inch TV market by 2009. Although LCDs are more expensive than PDPs, that prediction can be made mainly because of an advantage of LCD TVs over PDP TVs: LCD TVs can deliver full high definition (HD) resolution, see Table 1.1.

In order to increase profitability and market share, the LCD industry is speeding up efforts to develop new technologies. However, it has difficulty in ensuring price competitiveness because the material costs of LCD modules are higher than those of PDPs. To overcome this, researches are being conducted on technologies for (1) lowering costs of LCD panels by increasing the size of mother glass, (2) reducing costs of backlight units that account for about 40% of the total material costs of LCD modules and (3) improving quality.

The backlight unit is a light source for LCDs which are not self-luminous. It accounts for around 90% of the total power consumption and 40% of the material costs. As LCDs become larger and have higher resolution, backlight units are getting more important in both performance and cost-effectiveness. Currently, LCD TVs require brightness of at least $500 \, cd/m^2$. Since both larger screens and higher resolution decrease the transmittance of LCD panels, the light sources need to be brighter than $500 \, cd/m^2$ while consuming less power than competing display technologies.[2] In addition, research on advanced backlight driver technologies such as scanning, local dimming and field sequential technologies is being carried out to improve picture quality.[3] To meet such requirements, desirable backlight units of LCDs should have the following features: lower power consumption, higher brightness, wider color gamut, etc. This chapter describes basic components of – and requirements/specifications for – backlight units and reviews trends of backlight technologies.

Table 1.1 Comparison of resolution of LCDs and PDPTVs.

Size \ Type	LCD	PDP
30-inch range	HD 1366 × 768	
40-inch range	HD 1366 × 768	SD 640 × 480
	FHD 1920 × 1068	HD 1024 × 768
50-inch range	FHD 1920 × 1068	HD 1366 × 768
		FHD 1920 × 1080

1.2 Structure of LCD TV Backlights

Backlight units should satisfy the standards covering: (1) electrical characteristics such as power consumption, leakage current, energy efficiency and electromagnetic interference; (2) external appearance; (3) optical properties such as brightness, color uniformity and color gamut; (4) mechanical properties such as weight and thickness and (5) other aspects including reliability, mass productivity and quality of services.

Based on the location of the light source, backlights for LCDs can be divided into direct types and edge-lit types. Typically, direct types are used for TVs while edge-lit types are used for monitors and notebook computers.[4]

As shown in Figure 1.1, LCD TV modules of the direct type backlight units consist of multiple layers. Considered as a one-dimensional light source, lamps are placed linearly on the bottom chassis with reflective sheets. A diffuser plate and a diffuser sheet are mounted over the lamps to generate a uniform distribution of light, so that finally backlights can provide a two-dimensional distribution of light. In addition to spreading light from the lamps, the diffuser plate made of polymethylmethacrylate (PMMA) also functions as a support for holding up diffuser sheets. A prism sheet is used to increase the brightness measured normal to the surface after light is scattered through the diffuser sheet. Since the light passing through the prism sheet is not polarized, a reflective polarizer is added to recycle light lost to absorption, by controlling polarization and reflection.[5] Figure 1.2 shows how light distribution changes when light passes through each optical sheet used in a backlight unit. Typical backlight inverters are attached on the back of the bottom chassis, but recently those integrated with TV driving boards are also available in the market.

As stated above, optical sheets are needed to convert one-dimensional light sources into two-dimensional ones and to increase the brightness of LCDs. However, because they are expensive, research and development efforts are ongoing to minimize the use of optical sheets by merging multiple optical sheets into one or two sheets or by increasing the brightness of lamps. In particular prism sheets and reflective polarizer sheets, both of which are costly, need to be replaced or removed. Since they are also closely related to the performance of light sources, efforts are being made to improve existing light sources and develop new ones. In addition, the industry is conducting a number of activities to improve optical materials including the development of multifunctional sheets.

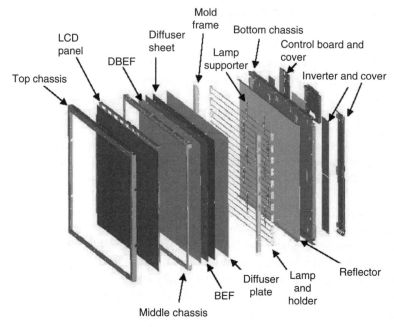

Figure 1.1 Structure of LCD TV modules.

1.3 Trends in LCD TV Backlights

Candidate light sources for backlights include cold cathode fluorescent lamps (CCFL), external electrode fluorescent lamps (EEFL), flat fluorescent lamps (FFL), light emitting diodes (LED), hot cathode fluorescent lamps (HCFL) and field emission lamps (FEL). Regardless of the type of light source used, currently some of the optical sheets mentioned above are chosen and combined depending on brightness and uniformity.

1.3.1 CCFL, EEFL, and HCFL

A cold cathode fluorescent lamp, used in most LCD TVs, has hollow cathode electrodes at both ends. The inside of a borosilicate glass tube with outside diameter of about 3–4 mm is coated with phosphors and a protective layer; then the tube is filled with neon, argon and mercury at 7–9 kPa (50–70 Torr). When a voltage is applied to the electrodes at the ends of the glass tube, electrons are accelerated by an electric field. When electrons collide with mercury vapor, they cause the vapor to emit ultraviolet light at 185 nm and

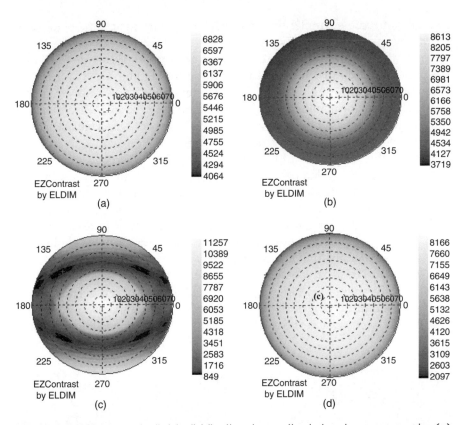

Figure 1.2 Changes in light distribution by optical sheet components: (a) diffuser plate, (b) diffuser plate/diffuser sheet, (c) diffuser plate/diffuser sheet/ prism sheet and (d) diffuser plate/diffuser sheet/prism sheet/reflective polarizer.

253.7 nm.[6] Phosphors absorb the ultraviolet light and glow. Almost all back-lights except those for cell phones use a CCFL. Compared with HCFLs, CCFLs have advantages of small diameter (up to 3–5 mm), high luminance and proven productivity and reliability.

However, it should be noted that for TV applications a certain number of lamps are needed to ensure high luminance and uniformity, depending on the size of the displays. Due to the negative resistance characteristic of plasma in discharge tubes, each lamp requires a transformer for over-current protection, leading to higher cost. Recently research is being conducted to drive several lamps simultaneously. Some companies have developed and commercialized the parallel circuitry which makes it possible to operate two to four lamps simultaneously.

As an alternative to the CCFL to reduce the backlight cost of LCD TVs, the EEFL has electrodes on the outside of the wall at both ends of the glass tube and operates using the wall charge generated when AC power is supplied. The basic principle of visible light emitted by the EEFL is almost the same as that of the CCFL. However, the glass bulb serves as a ballast capacitor and the electric current of negative-characteristic plasma can be controlled. In this sense, the EEFL can offer a major advantage in that it is possible to run all the lamps with just one or two transformers, resulting in cost reduction. In addition, the assembly process of EEFL-based backlights may be simplified by using a clip style, external electrodes give longer life and eventually the EEFL may be more reliable and easier to mass produce than the CCFL. Figure 1.3 shows the structural differences of backlights and inverters between the CCFL and the EEFL.

In spite of all these advantages, the EEFL has a serious problem of pinhole formation. As shown in Figure 1.4, high voltage and high current create pinholes around electrodes on the glass bulb.[7,8] Two mechanisms have been reported to be the cause of pinhole formation: dielectric breakdown due to

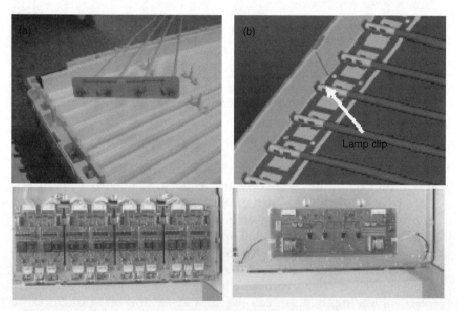

Figure 1.3 Structure of backlight and inverter of (a) CCFL and (b) EEFL.

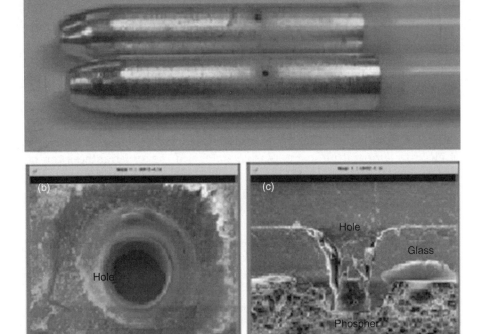

Figure 1.4 (a) Pinhole in an EEFL; (b) top view of pinhole; (c) cross-sectional view.

heat generation of electrodes and the following Joule heating and thermal breakdown due to simple thermal runaway.[9] The detailed mechanism is still unclear and further investigation is required. However, it is clear that a sufficient margin of operating power range can be obtained just by reducing the heat generation of electrodes, which can be easily achieved through careful electrode and system design.

It was expected that the EEFL would in due course replace the CCFL because of its several advantages at the early stage of development. However, the CCFL still keeps its position as the major light source for LCD backlight units. This has been possible through continuous cost reduction and performance improvement of the CCFL. For example, the two in one and/or four in one inverters using fewer transformers were introduced and alternative lamp assembly systems such as socket types were suggested. In other words, the CCFL is benchmarking the EEFL continuously. Furthermore, the EEFL has limitations in its applicable size. For large-size LCDs the operating voltage of the EEFL needs to be increased, but regulations limit the operating voltage of backlights to 2 kV to prevent ozone pollution. This may be overcome by increasing the area of electrodes if we neglect another potential problem of increases in bezel size.

To lower the cost, some LCD companies are trying to remove optical sheets by increasing lamp luminance. When the inner diameter of a lamp is decreased from 3 mm to 2 mm, its luminance increases due to the increase of electron temperature in the plasma state, and higher brightness can be achieved without the help of a prism sheet or reflective polarizer. However, UV and heat generation rate also increase and the increased flux of UV and heat may be harmful to several polymer materials, especially to polycarbonate (PC). Mold frames and reflective polarizer adhesives, mainly made of PC, show yellowing after long-term operation, decreasing reflectivity and transmittance and leading to defects. Several methods have been suggested to reduce yellowing: improving additives such as UV stabilizers, using UV cutting glass, coating lamp glass with TiO_2, etc..

Recently, Phillips and some other companies have tried to apply the HCFL of diameter under T5 (16 mm), typically used for general lighting, to LCD backlight. The HCFL has a very small cathode fall which is the main energy loss in glow discharge. Therefore it is more efficient and costs less than the CCFL. It was also reported that scanning does not decrease brightness, leading to better picture quality. However, current HCFL backlight systems have the disadvantage of generating a large amount of heat and their lives are shorter than those of CCFLs. Blackening of electrodes is another problem which results in the decrease in active area of panels and the increase in bezel size. To use the HCFL as a light source in LCD TV backlights, acceptable thermal management and the development of electrode materials and parallel operation technologies are priorities.

1.3.2 FFL

As a new approach, a two-dimensional light source has been studied for use in LCD TVs. The technology of a surface light source is intended to achieve further cost reduction by decreasing the number of sheets used in existing backlights. Unlike the CCFL and the EEFL shown in Figure 1.3, a flat fluorescent lamp is a two-dimensional built-in type light source as described in

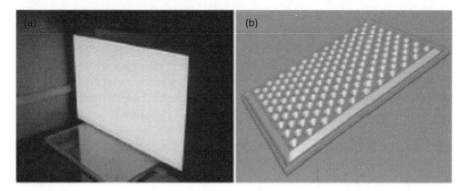

Figure 1.5 (a) 32-inch mercury discharge flat fluorescent lamp by Samsung Corning; (b) schematic diagram of a 30-inch mercury-free flat fluorescent lamp by OSRAM.

Figure 1.5; therefore, it may possibly increase luminous flux and achieve higher efficiency. It is also desirable from an assembly point of view because backlight manufacturing processes can be simplified and automated.

Surface light sources can be classified in various ways according to their shape and the gases used. This chapter focuses on those currently available in the market: mercury and non-mercury FFLs. In 1946, the first surface light source was introduced, using the serpentine structure and the structure filament terminals at both ends.[10] For life and efficiency improvement, there have been changes in electrodes such as hollow cathodes and external electrodes. A surface light source with the serpentine structure is advantageous: it simplifies manufacturing processes and reduces weight because the discharge space can be kept between the top and bottom molding glasses without a separate spacer. However, there are many shortcomings: it is hard to secure the technology related to the glass formation process; because electrodes are at the ends of the serpentine structure, a high voltage is necessary; sputtering of electrodes causes extra power consumption and leakage current; there may be non-emissive areas between channels and it is also hard to apply to large sizes.

As a result of recent studies, a 32-inch surface light source has been introduced.[11] Compared with a non-mercury lamp that uses xenon, a mercury lamp is a better choice because ultraviolet light is emitted better and it is converted into visible light more efficiently.

To prepare against environmental regulations about various pollutants in RoHS (Restriction of Hazardous Substances) which became effective in July

1, 2006, research efforts have been made to develop non-mercury lamps. Compared to mercury that provides longer wavelength (253.7 nm), xenon gas provides 147 nm and 173 nm. This difference in the wavelength of UV light causes low efficiency because the conversion losses are substantial. Also the higher ionization potential of Xe atoms results in energy loss due to elastic collisions between electrons and Xe atoms. In spite of these shortcomings, in 2004 Osram announced a 30-inch non-mercury surface light source that uses xenon discharges. Unlike PLANON I which has spacers between the top and bottom glasses, PLANON II has simplified the structure for manufacturing processes by introducing the reformation of the top glass and rearranging the electrodes with outside terminals.[12] Despite the superiority of the manufacturing technology itself, it has the disadvantages of huge heat due to the low efficiency (35 lm/W) and high power consumption. However, demand for non-mercury lamps keeps increasing and the development of non-mercury backlights along with LEDs is required.

To use FFLs in backlights for LCD TVs, there are several problems to solve. Improvement of lamp efficiency is a top priority. Without efficiency improvement, there is no way to remove optical sheets, resulting in no cost benefits. In the manufacturing process of FFL backlights, of course, cost reduction through the automation of assembly process can be promising. Because the material cost for the FFL is higher than that for the CCFL, it should be noted that the FFL's competitive power comes from efficiency improvement. Another important issue is the brightness stabilization time. It is normally required that the brightness level should rise up to 90% of the maximum brightness level within one minute after being turned on. In the case of surface light sources, there is a shortcoming that the brightness stabilization time at low temperatures (0 °C) tends to be increased because its heat capacity becomes bigger as the thickness and dimension of glasses increase. To compensate for this problem, high current needs to be applied at the early stage of the brightness stabilization time. However, in an external electrode type FFL which is under consideration for commercialization, high currents may cause a pinhole problem. Therefore, it is necessary to consider an advanced mechanical and structural design as well as material optimization.

1.3.3 LED

Since 2003, there has been a great deal of interest in LEDs as a next generation light source in backlights for TVs. An LED is a semiconductor device that converts electrical power to incoherent narrow-spectrum light. Normally, it uses III-V or II-VI compound semiconductors. Compared to other light sources, an LED has a longer life and requires a lower forward voltage. An LED has enough advantages to direct our attention to consider it as an advanced light source in backlighting. Its advantages include short response time, excellent color reproduction (>100%) if RGB LEDs are used, environmental-friendliness, durability and the degree of freedom for design.[13] Recently the efficiency of LEDs, which is a major drawback for adoption in the field, has greatly improved and the future of LEDs is becoming more promising. LEDs can be used in an LCD application in three ways: (1) white LED using blue LEDs with yellow phosphor, (2) white LED using UV LED and RGB phosphor, and (3) red, green and blue single-colored LEDs to make white. For LCD TVs, application of RGB single-colored LEDs is more favorable than the others because of its superiority in color reproduction.

In the usage of RGB single-colored LEDs as a cluster, there are many variations including: multiple chips in a package, RRGGB, RGGB and RGB combinations for a cluster. Small size LEDs require a larger number of LEDs. Basically, all these approaches are competing and coexist. It is difficult to forecast which one is going to be a winner, because the performance of LEDs is still advancing and cost effectiveness is improving very fast. The application of small size LEDs with a larger number of LEDs has its own advantages on color mixing and thickness reduction, more localization for local dimming, whereas it has disadvantages in packaging, assembly and cost effectiveness. Thus, the choice between small LEDs and large LEDs should be made based on a kind of trade-off between the cost and performance requirements, and also it is strongly dependent on the chip technology and other component technologies for LED backlight units.

So far, the advantage of the color reproduction of LED backlights has mainly been promoted for LCD TVs. By using RGB single-colored LEDs, the color gamut can be extended up to 105% or more compared to the CCFL backlights that have a color gamut of about 72% of NTSC standards. Because wide color expression which becomes possible by the color gamut extension can help reproduce true color, it may contribute to picture quality improvement. Since the selection of LEDs binning process is unavoidable to satisfy the requirements of the dominant wavelength and full width at half maximum (FWHM) for reliable color reproduction using LED, it becomes a critical factor that decreases the efficiency of the chip production and thereby increases cost. Besides, color gamut can be extended to 92% by improving the characteristics of phosphors for the CCFL and it can be expected to be improved further up to 100% soon. For this reason, the color reproduction as one of the merits of LED becomes less appealing. However, there is a kind of trade-off between color gamut and power consumption. In general, we need more electrical power as the color gamut increases. In other words, wide color gamut CCFLs consume more power than the normal CCFLs do. Therefore, as the efficiency of LEDs increases, LEDs can have more advertisement columns emphasizing lower power consumption.

It can be considered as a very strong point of LED backlights that a high contrast ratio and lower power consumption can be achieved by local dimming technology. By considering an LED as a point light source, the local dimming of LEDs increases the contrast ratio dramatically, which CCFLs and FFLs can do only slightly. However, for the local dimming, LEDs have to receive video signals in advance and feedback the converted signals into the backlight unit. So there are several things to be carefully considered such as the complex circuitry, appropriate radiation pattern of light and additional components.

Since a large number of LEDs, each of which is considered as a point light source in the LCD backlights, need to spread light to have enough uniformity and color mixing, we need some additional considerations to overcome such as color mixing and color shifting. Color shifting is color variations from one LED to another depending on the grouping of LEDs, and also from temperature dependency of LEDs. A more difficult part is that the temperature dependencies of LEDs differ from the colors of LEDs. However, the biggest hurdle is the higher cost caused by low efficiency, resulting in large number of LEDs to have the required brightness. To achieve comparable cost, there should be various efforts and trials such as higher chip yield rate, better materials and structures, higher efficiency, advanced LED packaging techniques, removal of redundant components for thermal management and elimination of optical sheets.

1.4 Requirements/Specifications for LCD TV Backlights

Table 1.2 shows the general specifications for the backlights in LCD TVs.

1.4.1 Luminance

LCD TVs in general require brightness of $500 \, \text{cd}/\text{m}^2$. Table 1.3 shows the brightness requirement for the backlight unit (BLU), assuming that the transmittance of the LCD panel is 3.5%. The brightness of a diffuser sheet and a reflective polarizer means that the brightness is measured at the top of the reflective polarizer. By considering polarization characteristics of the reflective polarizer, there is a gain of about two through its recycling when photons

Table 1.2   General specifications of LCD modules for TVs.

	Specification		Comments
	Year 2006	Year 2010	
Luminance (cd/m^2)	500	600	
Color reproduction (% NTSC)	72	100	
Correlated color temperature (K)	10 000	13 000	
Life (hour)	50 000	75 000	
Thermal management (max. °C)	50	40	
Power consumption (W)	150	70	40-inch

Table 1.3   Brightness of LCDTV backlights (requirement varies with panel transmittance).

	Diffuser sheet only	Diffuser sheet + DBEF
Brightness of BLU (cd/m^2)	14 000	7 000

travel to the LCD panel. Due to the increase in resolution of LCD TVs, for example, from HD (1366×768) to FHD (1920×1080), the aperture ratio drops by 15%. To compensate for the drop, the brightness requirements described in Table 1.2 need to be increased by an additional 15% to fulfill the specification of the full HD.

In the near future LCD TVs will require a brightness of 600 cd/m^2. This requirement can be met by improving the transmittance of the LCD panel and/or by increasing the brightness of backlights. In either case, a trade-off between cost and performance, an optimization of the power budget of the backlight and the development of advanced optical sheets to increase brightness should be considered.

1.4.2 Color Reproducibility

1.4.2.1 Improvement of the Color Reproducibility of the CCFL

Recently, Sony and Kasei Optonics have reported the improvement of the CCFL color gamut from 72% to 92% by applying a newly codeveloped phosphor. Figure 1.6 shows its emission spectrum compared with that of normal CCFLs. In the figure, the same blue phosphor is used. In the spectrum of red phosphors, the main emission peak shows 5 nm red-shift and the second

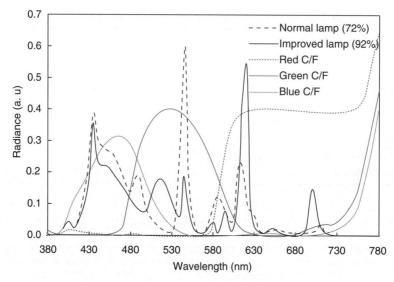

Figure 1.6 Comparison of transmission characteristics through color filter of CCFLs with 72% and 92% color reproducibility.

sub-peak of 586 nm disappears. In the case of green phosphors, the second sub-peak of 490 nm, which degrades the color gamut by transmitting both green and color filters, has been removed. As a result, the color gamut is enhanced by about 20% higher than that of normal phosphors. However, luminance and life of the wide color gamut CCFL are reduced by 25% and 40% respectively, compared to the normal phosphors. Additional improvement of the color gamut even to 100% is also possible if the color filter thickness and materials are further optimized. However, an improvement of the color gamut by increasing the color filter thickness may result in a decrease of panel transmittance, followed by high power consumption of the backlights.

1.4.2.2 Improvement of Color Reproducibility with LED

Emission spectra of the normal CCFLs and RGB LEDs are compared in Figure 1.7. Each LED shows a single main emission peak: red at 630 nm, green at 530 nm and blue at 450 nm. Showing of no other sub-peak, which is often observed in the spectrum of CCFL, indicates that the emission spectrum of the LEDs is well-matched with the absorption spectrum of the color

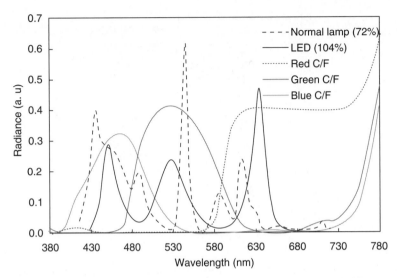

Figure 1.7 Transmission characteristic through color filter of a CCFL employing 72% color reproducibility and LED (104%).

filter. The color gamut of LEDs is about 104%, improved by over 30% higher than that of normal CCFLs. Figure 1.8 shows the enhancement of color gamut by introducing LED BLU in the CIE1931 color space. The color gamut of CCFLs is also shown for comparison.

1.4.3 Correlated Color Temperature

Table 1.4 describes the corresponding color coordinates on BLU and lamps in order to make the LCD module color temperature 10 000 K. Considering chromaticity changes from the lamp to the BLU, it is possible to notice that the change in the direction of y-chromaticity is greater than one in the x-chromaticity direction. It is due to more absorption of the short wavelength through the diffuser plates, diffuser sheets, prism sheets, and reflective polarizers. Considering the change in chromaticity from the BLU to the module, this can be explained from the optical characteristics of red and blue color filters of Figures 1.6 and 1.7. A blue color filter absorbs more photons of short wavelength, while a red one absorbs relatively fewer photons of long wavelength. In other words, because the transmittance of light of short

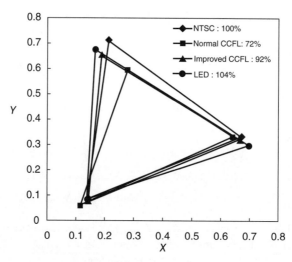

Figure 1.8 Comparison of color gamut of backlights adopting CCFLs with 72% and 92% color reproducibility and LEDs (104%).

Table 1.4 Color coordinates of lamps and backlights to satisfy the color coordinates of an LCD module.

	Module	BLU	Lamp
Color coordinate (CIE 1931)	(0.280, 0.290)	(0.265, 0.235)	(0.245, 0.205)

wavelength is lower than that of light of long wavelength, the chromaticity change in y-direction is greater than that in the x-direction.

As shown in Table 1.4, the corresponding chromaticity for the lamps leans to the bluish side. Therefore, the higher color temperature we need at the LCD panel, the lower is the efficiency of the lamps. For example, to set the color temperature of an LCD module to 13 000 K, the chromaticity coordinates for a lamp is approximately (0.235, 0.185). This makes a lamp very bluish, significantly lowering the lamp efficiency.

1.4.4 Thermal Management

Generally, the operating temperature of LCD TVs ranges from 0 °C to 50 °C and the temperature in liquid crystals also needs to be less than 65 °C. Considering that the phase transition temperature of liquid crystals is generally about 70 °C, it is obvious that the temperature margin is not enough. The temperature increase of the backlight unit is caused by both the driving of lamps and TV sets but mostly by the heat generated by lamps. Therefore, thermal management of the backlight units is essential for protection against heat. This can be achieved by improving lamp efficiency and mechanical structure.

1.5 Conclusions

Our discussion covers the requirements and technology trends for various kinds of light sources in backlights for LCD TVs. So far the CCFL has dominated the backlight market, while today's LCD backlight market shows transition to the next generation with advanced light source technologies including EEFL, FFL and LED. This is the result from the LCD industry's continuous development efforts to gain an advantage over competing technologies such as PDPs. The CCFL is expected to keep its position for some time. EEFLs and FFLs are claiming their own advantages and will maintain their position with a certain portion of the market. However, they are not enough for a paradigm shift of light source technology. As a light source in the backlight unit, LEDs may become more popular and can be a potential choice to offer the paradigm shift. The advantages of the LEDs have been promising and also many companies are investigating their possibilities. However, for real implementation, further cost reduction of the LEDs is necessary. In the high-end products with small volume, an early market entry of the LED is possible. To position LEDs as the mainstream light source in LCD TVs, LED manufacturers and customers still need to wait till LEDs arrive at prices comparable to CCFLs.

In conclusion, it is expected that CCFLs and LEDs will be the major competing technologies for now and in the future. The key factors are: how fast these two different light sources can absorb technical advantages from each other, how to improve their performance and how to optimize the relationship between price and performance. Such an effort can become meaningful when LCDs have a better position than other flat panel displays in the large-size TV market.

References

[1] Yoshida, Y. *et al.* (2004) *Proc. 24th Intl Display Res. Conf.*, pp. 30–33.
[2] Mikoshiba, S. (2005) Crystal Valley Conference on LCD Backlights, pp. 3–4.

[3] Oh, E. Y. *et al.* (2004) *Proc. 24th Intl Display Res. Conf.*, pp. 799–802.

[4] Anandan, M. (2001) *SID '01 Seminar Lecture Notes*, F2/37.

[5] Kim, J. H. et al. (2004) *Proc. 24th Intl Display Res. Conf.*, pp. 795–798.

[6] Noguchi, H. and Yano, H. (2004) *SID Intl Symp. Digest Tech. Papers*, pp. 935–938.

[7] Cho, G. S. *et al.* (2004) *J. Phys. D: Appl. Phys.*, **37**, p. 2863.

[8] Guzowski, L. T. (2002) US Patent No. 6 341 879.

[9] Waymouth, J. F. (1971) *Electric Discharge Lamp*, MIT Press.

[10] Holmes, J. E. (1946) US Patent No. 2 406, p. 146.

[11] Hwang, I. S. *et al.* (2004) *SID Intl Symp. Digest Tech. Papers*, pp. 1326–1329.

[12] Ilmer, M. *et al.* (2000) *SID Intl Symp. Digest Tech. Papers*, pp. 931–933.

[13] West, R. S. *et al.* (2003) *SID Intl Symp. Digest Tech. Papers*, pp. 1262–1265.

2

Improvement of Moving Picture Quality by Means of Backlight Control

T. Yamamoto

Hitachi, Ltd

2.1 Introduction

Along with the rapid growth of the LC-TV market, there are more and more chances for us to look at moving images which are displayed on LCDs. Picture quality of the moving images of the present-day LC-TVs, however, is not satisfactory. For example, when watching a ball game, the balls tend to blur.[1] This is caused by the relatively slow response of LC materials, and also by the 'hold-type' image expression scheme. The slow liquid crystal response has been pointed out for a long time. The adverse effect of the hold-type display, however, has only been known for a few years and the mechanism of the motion blur is being extensively investigated. This chapter explains the origin of the blur, its measurement methods and techniques for reducing the blur. Special emphasis is given to an explanation of backlight blinking.

LCD Backlights   Edited by Shunsuke Kobayashi, Shigeo Mikoshiba and Sungkyoo Lim
© 2009 John Wiley & Sons, Ltd.

2.2 Blur of Moving Images on LC Displays

2.2.1 Origin of Motion Blur

An image of a moving object displayed on an LCD looks different from an image shown on a photo taken with a still camera. Blur sometimes appears at the edge of the moving object. This is because 'an image of a pixel' moves on the retina while emitting light. In LCDs the light emission generally continues for a TV frame, thus called 'a hold-type' emission, as explained in Figure 2.1. The vertical axis represents emission intensity and the horizontal axis is time. A pixel of a CRT emits light impulsively (about 1 μs) only once in a TV frame time of 16.7 ms (Figure 2.1(a)). A pixel of an LCD, on the other hand, holds the light emission during the TV frame (Figure 2.1(b)). Liquid crystal has a relatively slow response time, which is indicated by dotted lines in the figure. CRT phosphor also has decay times, but much faster than the response of liquid crystal materials.

The relationship between the signal applied to an LCD and an image perceived by human eyes is explained in Figure 2.2. For simplicity, the response of the liquid crystal is assumed to be zero. The signal applied to the LCD consists of dark pixels (indicated in black) which moves to the right at a speed of four pixels per TV frame in a white background. The LC display rewrites the image once in each TV frame of 60 Hz, or every 16.7 ms. Once written, the image is kept unchanged during the TV frame. Therefore the displayed image on the screen becomes stepwise as shown in Figure 2.2.

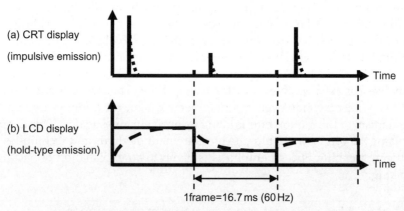

Figure 2.1 Light emission schemes of CRT and LCD.

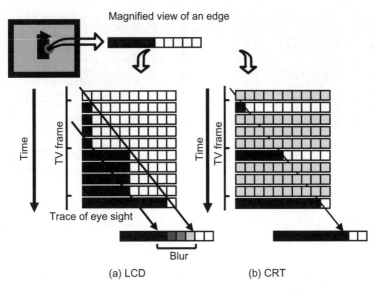

Figure 2.2   Generation of blur in motional LCD images.

The human eye has the following two characteristics of perception. One is that, when a moving object on a display screen is observed, the eyeball follows the image 'smoothly'.[2] That is, if the object changes its speed of motion within several TV frames, then the eyeballs follow the object at an average speed of the object. Another is that eyes have an after-image effect with which light emission is integrated for a time period of roughly 30 ms. In Figure 2.2(a), the eyeball follows slanted arrows, although the displayed image has a stepwise motion. The perceived image is found by integrating the light input to the retina along the slanted arrows. A resultant image perceived by the eye is shown at the bottom of the diagram. The image has a blurred zone between the white and black regions.

A CRT refreshes the displayed image every TV frame, just like LCDs. The light emission, however, decays rapidly and there is no more emission until the next addressing. Therefore, even if the eye traces a slanted path on the screen, no blur appears as shown in Figure 2.2(b). If one further considers the response time of the LC material, then the blur becomes wider. Figure 2.3(a) is when the response is zero and Figure 2.3(b) is when the response is about one TV frame (16.7 ms). The blur becomes wider by the amount corresponding to the response time.

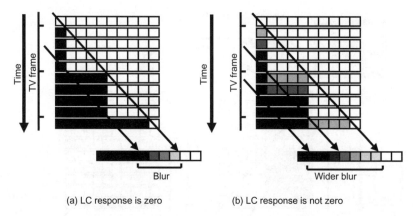

(a) LC response is zero   (b) LC response is not zero

Figure 2.3 LC response and widening of blur in motional LCD image.

2.2.2 Picture Quality Evaluation Methods of Moving Images: N-BET and MPRT

When reproducing perceived images with motion blur on a hard copy, an evaluation system is required in which characteristics of the eye are incorporated. The system includes a time-based image integration measurement system,[3] a pursuit camera system,[4] and a rotating mirror system.[5] The first two systems are shown in Figure 2.4. For the time-based image integration measurement system, a scrolling image is displayed on an LCD whose picture quality is to be examined. Then still images are captured with a CCD camera at a constant time interval in step with the original image. Finally the intensities of the captured images are integrated by moving them along the direction of motion of the eye. This reproduces the image perceived by human eyes. In the pursuit camera system, a camera rotates so that it pursues the image on an LCD screen, simulating the motion of the eyeball. In the rotating mirror system (not shown in Figure 2.4), a galvano mirror rotates to follow the image motion.

The degree of blur of the reproduced image can be evaluated by plotting a luminance profile. BEW (blurred edge width) of Figure 2.4 quantifies the degree of blur by defining the width within which luminance varies from 10% to 90% of the peak value.[6]

It has been found experimentally and theoretically that the BEW is proportional to a scroll speed.[3] Considering this, N-BEW (normalized BEW), which is BEW normalized by the scroll speed, is proposed. Although BEW depends on the speed of the sample image, N-BEW does not. However,

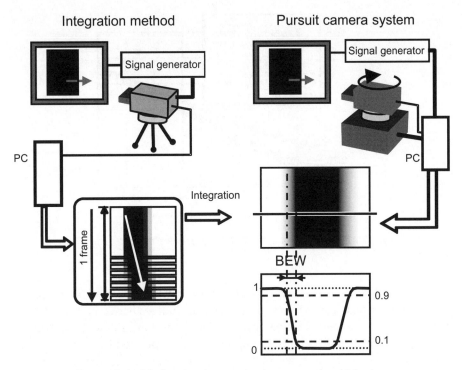

Figure 2.4 Methods of reproducing perceived blurring.

N-BEW still depends on the frame time, giving different results, for example NTSC and PAL. Accordingly N-BEW is again normalized by the frame frequency to obtain N-BET (normalized blurred edge time). Once again N-BEW and N-BET are expressed by Equations (2.1) and (2.2) as:

$$\text{N-BEW (frame)} = \text{BEW (pixel)}/\text{scroll speed (pixel/frame)} \quad (2.1)$$

$$\text{N-BET (s)} = \text{N-BEW (frame)}/\text{frame frequency (frame/s)}. \quad (2.2)$$

Experimental and simulated values of the N-BET for various LCDs are shown in Figure 2.5 as a function of liquid crystal response time. When the frame frequency for vertically scanning a frame, f_v, is 60 Hz and also when the response time is longer than 10 ms, N-BET increases linearly with the LC response time. However, if the response time is shorter than 5 ms, N-BET becomes independent of the response time. Here, the blur originates only

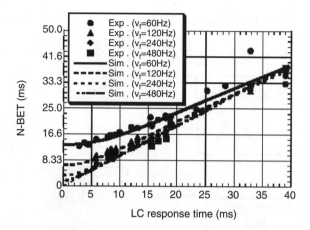

Figure 2.5 LC response time versusN-BET.

from the hold-type display scheme. The value of N-BET defined by Equation (2.2) corresponds to cases in which luminance changes between 0 and 100%. The response time between one gray level and another gray level, however, is longer. MPRT (moving picture response time) is defined by an average of seven N-BET values measured between equally spaced gray levels.[7] Evaluation methods similar to N-BET and MPRT were standardized by the Video Electronics Standard Association (VESA) as FPDM2UPDATE in May 2005.

2.3 Methods of Reducing Motion Blur

As explained above, the motion blur is caused by an adoption of the hold-type image expression technique. Methods of reducing the blur are suggested by the displays that do not show the blur. An object illuminated by sunlight is emitting light constantly, in a fashion similar to the hold type, but no blur appears to the image even when the object is moving. Light emission from CRT pixels is impulsive, and also no blur is observed. The difference between the object in the sunlight and an image displayed on an LCD lies in the difference of refresh rate of the images. Namely the object in the sunlight is continuously refreshing the output luminance along with the motion, but LCDs refresh the output luminance only once in a TV frame.

From these considerations the use of a higher frame frequency, for example doubling the frame frequency from 60 Hz to 120 Hz (as shown in Figure 2.6), can be suggested. In order to adopt the high-frequency drive technology,

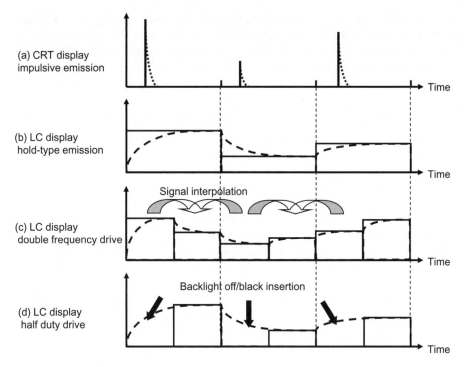

Figure 2.6   Reduction of blur for hold-type emission LCD.

however, various issues have to be solved. One of these is that the LCD has to be driven at a speed twice that of the 60 Hz drive. Since the drive voltage waveforms applied to the LCD panels are deformed even with the 60 Hz drive due to the load of electrode wirings especially for large screens, an increase of the drive speed results in degraded image expression. For high resolution displays with a larger number of addressing pixels, turn-on time of a–Si transistors as well as response time of the liquid crystal materials are already close to the limiting values even for the 60 Hz drive, making the 120 Hz drive even more difficult. One of the sophisticated techniques which has to be implemented here is to create 120 Hz images in a TV receiver from 60 Hz NTSC images.

A simpler, yet effective method of reducing the blur can be deduced from the impulsive light emission of CRTs. The method shortens the light emission period during a TV frame time – a low-duty drive. For instance an image is expressed only during the former half of the TV frame, while nothing is expressed in the latter half. In order to realize this, a black signal can be applied to the latter half (black insertion). Alternatively, the backlight can be turned off during the latter half of the TV frame.[8] A shortcoming of the technique is that the luminance is reduced. For the half-duty drive, for instance, the output luminance is halved. The luminance reduction can be compensated for by increasing the backlight output flux. With the black insertion technique, the contrast of the displayed image is degraded since the black image cannot be made perfectly black due to leakage of light through the pixels in an off state.

Validity of the techniques introduced above can be assured by studying motion blur of the simple image of Figure 2.2. Figure 2.7(a) is for the double frequency drive, and Figure 2.7(b) is for the black insertion with a duty factor of a half. It can be understood that the blur width is reduced to a half for both cases. The double frequency technique and the half-duty technique have an identical effect in reducing the blur. Suppose that we express only the first half of the images during the TV frames, while the second half images are changed to 'black' in the double frequency drive. Then the resultant image becomes identical to that of the black insertion. During the black insertion period, the human eyes 'unconsciously create' images which connect the images before and after the black insertion smoothly. The difference between these two techniques is that the half-duty method reduces the luminance by a factor of two.

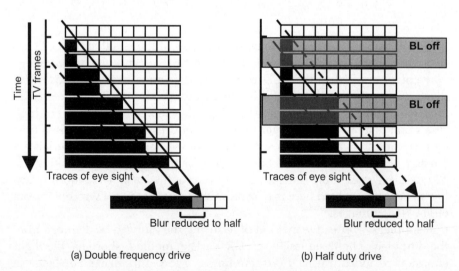

(a) Double frequency drive   (b) Half duty drive

Figure 2.7 Two methods of reducing blur.

From the above discussions, n-time frame frequency drive and $1/n$ duty drive are equivalent in terms of the blur reduction. Values of N-BET in Figure 2.5, when the liquid crystal response time is zero, indicate that the N-BET is inversely proportional to n. When the response time of liquid crystal is not zero, N-BET is not reduced to $1/n$. These techniques become less effective as the LC response becomes slower, and become totally ineffective when the response is 30 ms or longer.

Generally the n-time frame frequency drive and $1/n$ duty drive do not give identical results. Figure 2.8(a) is the double frequency drive in which the blur is not significantly reduced compared to the case of Figure 2.3(b) due to the slow liquid crystal response. On the other hand, the blur is reduced appreciably in the half-duty drive of Figure 2.8(b) even with the slow response. This is because the backlight is turned off in time with the switching of the LC. The initial change of the LC transmission, therefore, is hidden by the non-emission of the backlight. It should be noted that the build-up time of the backlight intensity is assumed to be fast enough in the figure, which is true for LEDs. For CCFLs, however, the slow response of phosphor of the order of ms extends the blur width. This, again, can be improved by synchronizing the build-up timings of the LC and backlight.[8] It therefore is possible to reduce the blur more with the $1/n$ duty drive than with the n-time frame frequency drive.

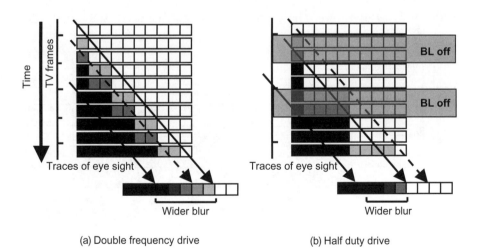

(a) Double frequency drive   (b) Half duty drive

Figure 2.8   Reduction of blur considering LC response.

2.4 Backlight Blinking

The $1/n$ duty drive introduced above will be examined further. Synchronization of the build-ups of the LC and backlight is essential for the reduction of the blur. Since LC-TVs are addressed with a one-horizontal-line-at-a-time scheme, the timings of the LC build-up depend on the vertical position of the display. Figure 2.9 explains a case when all the CCFLs of the backlight unit are turned on/off simultaneously. The scanning of the LC panel in the upper, central and lower regions are represented in (a), (b) and (c) of Figure 2.9. It can be seen in the diagram that the upper and lower regions are associated with ghosts, or thin edges. By dividing the BLU into several sections and activating them in turn in a scrolling manner while synchronizing with the LC addressing time as shown in Figure 2.10, the ghosts become less visible. In the scrolling backlight with a larger number of divisions, however, an extra cost of the CCFL drivers is assessed.

Figure 2.11 summarizes the methods of reducing the motion blur. Both reduction of the hold-type artifacts and improvement of the LC response are necessary. The improvement of the LC response to faster than 10 ms, however, results in only fractional improvement of the picture quality. Making the response 5 ms or faster has no effect at all on the blur width. The commercially available LCs have a response of 10 ms or less, indicating that the further improvement of the LC response does not reduce the blur much. The reduction of the hold-type artifacts can be made by the n-time frequency drive or $1/n$ duty drive. The n-time frequency drive needs data interpola-

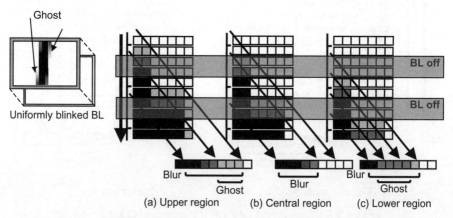

Figure 2.9   Blur with uniform blinking of BL.

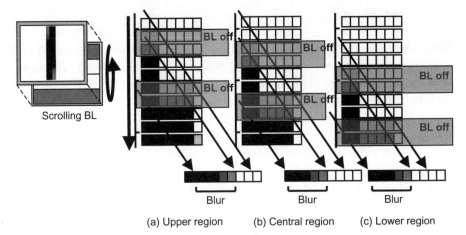

(a) Upper region   (b) Central region   (c) Lower region

Figure 2.10   Blur with scrolling BL.

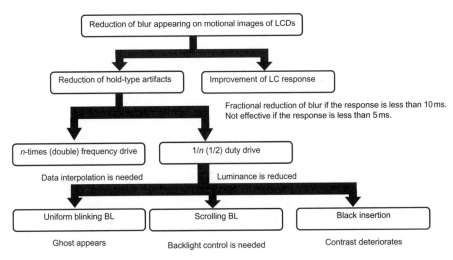

Figure 2.11   Methods of reducing blur.

tion, which requires a substantial amount of signal processing. The $1/n$ duty drive can be realized by blinking the backlight, by black insertion, or by adopting both of these. There are two types of the blinking backlight: uniform blinking and scrolling. The blinking backlight gives higher contrast compared with the black insertion, although a cost increase is associated with it.

2.5 Conclusions

The generation mechanism, measurement and improvement of LCD motion artifacts were explained. Standardizations of N-BET and MPRT are being considered. These evaluation techniques will help improve the picture quality of moving images.

References

[1] Kurita, T. *et al.* (2001) *SID'01 Digest*, pp. 986–989.
[2] Yee, H. *et al.* (2001) *ACM Trans. Computer Graphics*, **20**, 39–65.
[3] Yamamoto, T. *et al.* (2000) *SID '00 Digest*, pp. 456–459.
[4] Oda, K. *et al.* (2002) *Proc. EURODISPLAY '02*, pp. 115–118.
[5] Oka, K. and Enami, Y. (2004) *ITE Journal*, **58**, 1248–1253.
[6] Yamamoto, T. *et al.* (2002) *Proc. IDW '02*, pp. 1423–1424.
[7] Nakamura, Y. (2003) *Proc. IDW '03*, pp. 1479–1482.
[8] Furuhashi, T. *et al.* (2002) *SID '02 Digest*, pp. 1284–1287.

3

Multiple Primary Color Backlights

H. Sugiura

Mitsubishi Electric Corporation

3.1 Present Status

More than nine years have passed since the International Electrotechnical Commission (IEC) published an international standard IEC 61966-2-2 on sRGB, which is widely used as a standard color space.[1] When standardization activities of sRGB started in the middle of the 1990s, the major color graphics monitors were CRTs, and hence the sRGB was made considering the CRT performances. LC displays, at that time, were still at a developmental stage, trying to reproduce colors of CRTs. Around 2000 and after, the color reproduction capability of LCDs became comparable to that of CRTs, allowing LCDs to enter the color graphics monitor business. These two devices are now capable of reproducing colors close to the 'reference image display system characteristics' which is defined in sRGB.

Thus color reproduction of various devices is approaching the sRGB limit, and various issues on color matching between monitors are disappearing. The color matching between the monitors and other types of devices, however, is still incomplete. For instance, some of the colors of ink jet printers cannot be reproduced by monitors. This is due not to the monitors

LCD Backlights   Edited by Shunsuke Kobayashi, Shigeo Mikoshiba and Sungkyoo Lim
© 2009 John Wiley & Sons, Ltd.

Table 3.1 Standardizations of extended color spaces.

Document number, title	Present status
IEC 61966-2-1 Amd1: sYCC, bg-sRGB[2]	Published in January 2003
IEC 61966-2-2: scRGB[3]	Published in January 2003
JEITA DCF 2.0, Exif 2.21: DCF optional color space[4]	Published in September 2003
IEC 61966-2-4: xvYCC[5]	Published in January 2006
IEC 61966-2-5: opRGB[6]	New Work Item Proposal approved in January 2005

but to the color management of sRGB. Namely, an ink jet printer can reproduce colors which are outside of the sRGB color space. In order to fully utilize the colors that ink jet printers can reproduce, various color spaces which are wider than the sRGB have been proposed,[2–6] and some of them have been published as international standards. Table 3.1 lists standardization activities of the extended color spaces. sYCC[2] expresses image data by using both luminance signals and color difference signals. When transformed to RGB signal, components smaller than 0% and larger than 100% can be expressed. This indicates that, although sRGB coordinates of primary colors are utilized, sYCC can express colors of a wider space. Hence sYCC is widely used under the name 'Exif Print'. 'DCF optional color space' of DCF 2.0, Exif 2.21 has been standardized by the Japan Electronics and Information Technology Industries Association (JEITA)[4] and is now being discussed by IEC as IEC 61966-2-5 opRGB.[6]

sRGB denotes a color space defined by output devices (display reference color space). scRGB[3] of Table 3.1, on the other hand, denotes a color space defined by input devices (relative scene RGB color space). xvYCC[5] or extended-gamut YCC color space is an advanced version of sYCC for video applications. This completely covers the 'Munsell Color Cascade' of surface colors.

Along with advancements of standardization of various color spaces, images with improved colors appeared. For instance, sYCC images having a wider color gamut than sRGB can be reproduced easily with a digital still camera which adopts Exif 2.2. Even home use ink jet printers can reproduce images having a color gamut wider than sRGB. Colors reproducible even with these commercially available digital still cameras and ink jet printers cannot be reproduced on CRT and LCD monitors. Thus it is necessary to widen the color gamut of monitors.

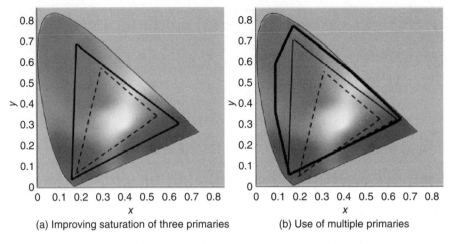

(a) Improving saturation of three primaries   (b) Use of multiple primaries

Figure 3.1 Methods of widening color gamut.

Methods of widening the color gamut of the CRTs and LCDs are:

(a) improving color saturation of the three primaries
    (1) CRTs with new phosphor,[7]
    (2) improvement of CCFL and color filters,
    (3) use of LED backlight;[8]
(b) using multiple (more than four) primaries
    (1) multiple primary color filters,[9]
    (2) multiple primary emission from backlight,[10]
    (3) a combination of these two.[11]

Items (a) and (b) listed above are explained in Figures 3.1(a) and (b), respectively. (a) is to widen the area of the triangle by improving the positions of the apexes. Several display devices of this type have already been commercialized. (b) is to widen the area by providing four or more apexes. Mitsubishi Electric recently made a 23-inch WXGA LCD monitor[10] which employs six primary colors from an LED backlight unit. The monitor provides the widest color gamut in the world.

3.2 Technological Impacts

The 23-inch six-primary-color WXGA LCD monitor can express images with a color gamut of 175% of the conventional sRGB on the CIE xy chromaticity diagram. With the wide color gamut, more than 95% of the realistic surface

Table 3.2   Specifications of a prototype LC monitor with six-primary LED backlight.

Item	Specification		
LC panel	OCB mode color TFT LC panel (normally white)		
Active display area [mm]	501.12(H)		
	300.67(V)		
	(diagonal: 23-inch)		
Number of pixels	1280(H)		
	768(V)		
Pixel pitch [mm]	0.3915(H)		
	0.3925(V)		
Number of displayed colors	16 777 216		
Peak luminance [cd/m^2]	80		
Video signal interface	TMDS		
Light source of BLU	Six-primary, power LEDs		
Color gamut (CIE xy color coordinates)		x	y
	R1	0.682126	0.307545
	R2	0.664004	0.321485
	G2	0.290582	0.665554
	G1	0.130854	0.580307
	B1	0.111746	0.173119
	B2	0.153561	0.060425
	Reproducible area of color (versus. sRGB) 175 [%]		

colors (eliminating colors emitted from lasers, for example) can be expressed. There were several presentations on multiple-primary color monitors at the 2005 SID (Society for Information Display) Symposium. The present 23-inch WXGA LCD monitor attained the widest color gamut of all the direct-view display monitors. Most of the hard copy printers use four-color YMCK (yellow, magenta, cyan and black) offset printing. Recently a standardization group added more colors for hard copies. The prototype LCD monitor developed here also has the aim of reproducing all these surface colors. Therefore the monitor may be used for soft-proofing of high-end devices. Specifications of the monitor are listed in Table 3.2.

3.3 Operation of Prototype, Six-primary-color Monitor

3.3.1 Video Input Processing

Figure 3.2 shows a block diagram of the prototype monitor driver. The video signal comes through two digital visual interface (DVI) connectors.[12]

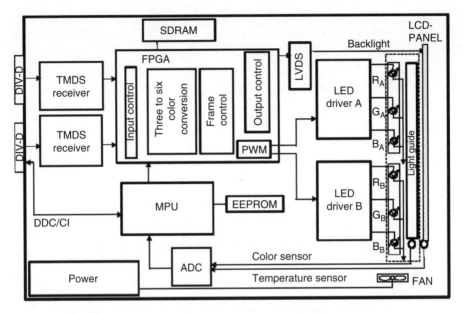

Figure 3.2   Block diagram of prototype monitor driver.

The circuit allowsfor the selection of either of the two 8-bit digital RGB video inputs. The two inputs can alternatively be combined in the FPGA (field programmable gate array) for 10-bit operation, assigning the upper 8 bits and lower 2 bits to these inputs respectively.

3.3.2 Frame Rate Control

The frame rate and dot-clock of the video signal are transformed into a frame rate and a dot-clock of LCD drivers in the FPGA.

3.3.3 Video Output and Driving of LEDs

Three primary color LEDs belonging to group A and another three belonging to group B are ignited in turn following the sequence of Figure 3.3. Synchronously with the LED emission, two sets of video signals for the respective groups A and B are applied to the LC panel. Figure 3.4 explains the control of backlights and video signals. For the LCD, the optically compensated bend (OCB) mode was used. Although the OCD has a relatively high response speed, it is not high enough, resulting in contrast degradation and color shift due to cross-talk between the subframe signals. In order to

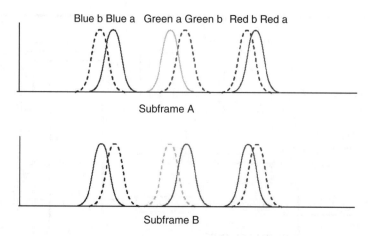

Subframe A

Subframe B

Figure 3.3 Ignition sequence of LED backlights.

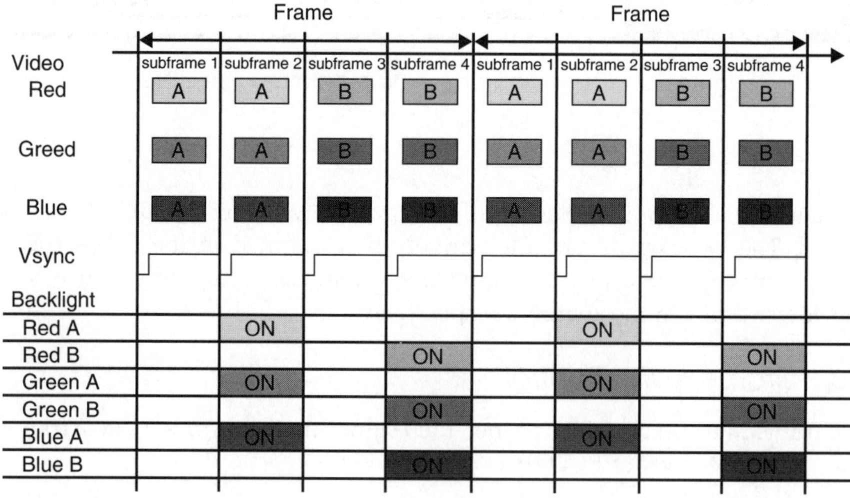

Figure 3.4 Control of video and backlight (LEDs) for multi-primary color.

eliminate these, address timing and light emission timing are separated. A frame is divided into four subframes 1–4 as indicated in Figure 3.4. In subframes 1 and 2, videos corresponding to group A are repeated. In order to eliminate artifacts arising from the slow LC response, group A backlights (Red A, Green A, and Blue A) are turned on only during the subframe 2. Likewise in subframes 3 and 4, videos corresponding to group B are repeated, while group B backlights (Red B, Green B, and Blue B) are turned on only during subframe 4.

3.3.4 Optical Feedback Circuit

Intensity and wavelength of LED output depends on drive current. Therefore the output intensity of LEDs is controlled by using the pulse width modulation (PWM) technique, while keeping the drive current constant. White-balanced luminance and color of groups A and B are made identical so that flickering can be eliminated. As shown in Figure 3.5, output intensity and wavelength also depend on the operating temperature of LEDs. Figure 3.6 depicts the optical and temperature feedback control circuit. A photo sensor (Si photodiode) attached to the module feeds back the intensity and color data to the comparator. Also a temperature sensor sends data to compensate for the shift of intensity and wavelength, as well as for variations of transmission and color of LC panels.

3.3.5 Communication Control

The prototype monitor is equipped with a display data channel command interface (DDC/CI)[13] for communicating with a host computer. Luminance, color and color transform can be controlled from the host computer. Also by combining with the color sensor, white point, gamma characteristics and hue can be automatically adjusted. When controlling LED luminance and color coordinates by using PWM, 16 bits are assigned to each color. A virtual

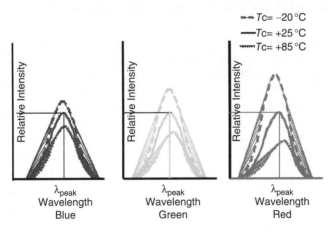

Figure 3.5 Dependence of intensity and wavelength on LED temperature.

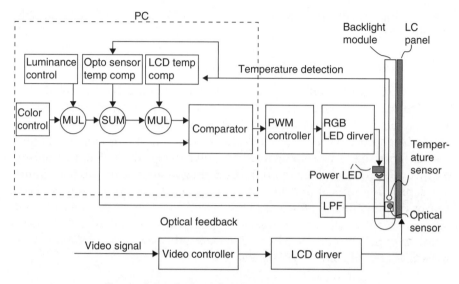

Figure 3.6 Optical feedback control circuit.

control panel (VCP)[14] is responsible for the operation with the support of a DDI/CI. A specially developed version of VCP is used in addition to the VCP which has been standardized by VESA monitor control command sets (MCCS).[14]

3.4 Details of a Six-primary-color Backlight Unit

The prototype monitor uses six kinds of LEDs. As mentioned above, these are divided into two groups A and B having different trios of RGB, which are turned on sequentially. This consists of a 'field sequential backlight system': LEDs of the groups A and B produce 'white' having different spectroscopic characteristics. With the help of color filters, six primary colors are expressed at different positions and timings.

Figures 3.7(a) and (b) show the LCD transmission spectra and LED emission spectra of groups A and B. The curves denoted as CF-R, CF-G, and CF-B indicate transmission spectra of the LC panel with color filters. Ra, Ga and Ba are the emission spectra of group A LEDs, and Rb, Gb and Bb are those of group B LEDs. Note the differences in the emission spectra of groups A and B. Chromaticity coordinates of displayed colors of groups A and B as observed on the LCD screen are plotted on the CIE diagram of Figure 3.8.

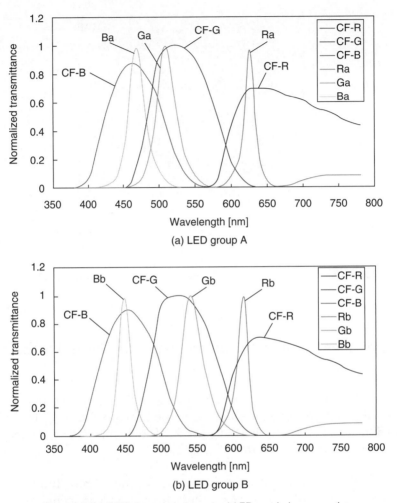

Figure 3.7 LCD transmission and LED emission spectra.

The six primaries are first mixed in the backlight unit, and then decomposed by the color filters. Generally the transmission spectra of color filters are wider than the emission spectra of LEDs so that color contamination occurs. In the prototype monitor, LEDs were selected so that most of the colors defined by the 'Munsell Color Cascade', as well as the colors of ink jet printers, could be reproduced. The color gamut of the monitor will be described later.

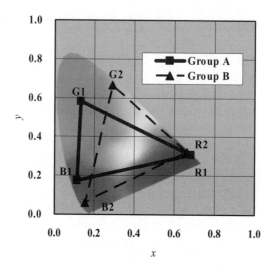

Figure 3.8 Coordinates of displayed colors for two groups of LEDs.

3.5 Signal Processing of Transforming from Three Primaries to Six Primaries

In the prototype monitor, image data corresponding to three primaries, Ri, Gi and Bi, are transformed into six primaries, Ra, Ga, Ba, Rb, Gb and Bb, following the signal processing of Figure 3.9. The circuit operates in a manner similar to a natural color matrix (NCM) color transform.[15–19]

Step 1: The original RGB image data are separated into chromatic and achromatic components. Here, the achromatic component consists of black-gray-white data without hue. The chromatic component is the subtraction of the achromatic component from the original data.

Step 2: Six primary component data and inter-primary component data are created from the chromatic component. The primary component consists of red, yellow, green, cyan, blue and magenta. The inter-primary component is a color between these primary colors.

Step 3: From the primary component data, inter-primary component data and achromatic component data, Ra, Ga, Ba, Rb, Gb and Bb are obtained.

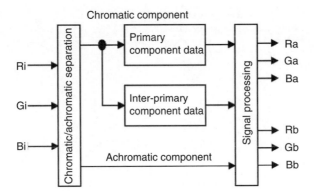

Figure 3.9 Transform from three primaries to six primaries.

3.6 Color Gamut of the Prototype Monitor

The color reproduction areas of a six-primary-LED backlight, a three-primary-LED backlight and an sRGB are compared in Figure 3.10(a). For simplicity, a two-dimensional CIE xy chromaticity coordinate is used here. Precise comparisons of the color gamuts, however, need three-dimensional considerations using, for example, CIELAB color space. Figure 3.10(b) compares the present three/six-primary LED backlights with SWOP (Specification for Web Offset Publications) which defines the standard colors used by US offset printers. From the figure it can be concluded that, if coverage of SWOP is the main objective, that is soft-proofing of YMCK offset printing for desktop publishing is the main objective, then the three-primary LED backlight is satisfactory.

The Munsell Color Cascade,[20] which is widely used for evaluating a color reproduction area, is compared in Figure 3.10(c). It can be seen that neither the conventional sRGB, nor the three-primary LED backlight unit, is satisfactory in representing the Munsell Color Cascade, especially for highly saturated cyan (or emerald green). The six-primary monitor, on the other hand, covers almost all the surface colors. The coverage ratios of the Munsell Color Cascade for various types are listed in Table 3.3, in which three-dimensional calculations were performed. The color reproduction area of ink jet printers is compared with that of the present monitor in Figure 3.10(d). The monitor includes almost all the colors.

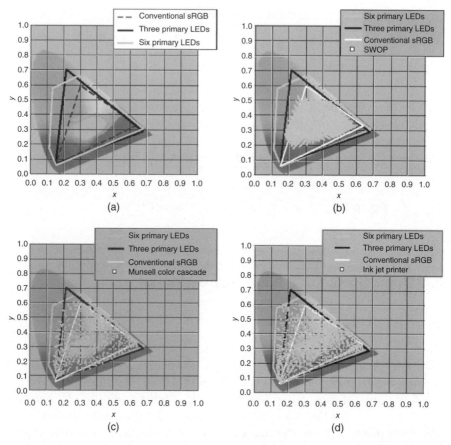

Figure 3.10 (a) Color gamut comparison; (b) color gamut comparison with SWOP (c) color gamut comparison with surface color; (d) color gamut comparison with ink jet printer.

Table 3.3 Coverage ratio for Munsell Color Cascade.

Type	Coverage ratio (%)
Six-primary LED backlight	95.58
Three-primary LED backlight	80.62
Conventional sRGB	52.67

3.7 Other Techniques for Multiple Primary Color LC-TVs

3.7.1 Three Primary LED Backlight with Two Color Filters[11]

The method employs a low pass and a high pass filters, together with three-primary LEDs. Color mixing is done spectrum sequentially by using two TV frames. The panel has a high aperture ratio, yielding high efficacy. Also it provides a wide color gamut. There is less flickering and less color separation compared to the conventional field sequential color drive which uses three frames instead of two. The device is adequate for mobile displays.

3.7.2 Alternating Ignition of Two Sets of CCFLs[21]

Two sets of CCFLs having different emission spectra are turned on sequentially. Combined with conventional RGB color filters, six colors are generated. Due to the limited selection of phosphor particles, improvement of the color gamut is not significant.

3.7.3 Six Primary Color Filters[9]

The LCD utilizes six primary color filters, RGBCMY. A completely new panel design is required, resulting in higher cost. Figure 3.11 compares the color gamut of these techniques, and Table 3.4 summarizes the areas of color reproduction on the CIE xy chromaticity diagram. In the table, the area of sRGB was made equal to 100. It can readily be found that the present six-primary LED monitor yields the highest color reproduction capability.

3.8 Remaining Issues

A liquid crystal display with the six-primary LED backlight unit can express colors which cannot be obtained with the three-primary LCD backlight unit. One of the most important applications is soft-proofing for high-end desktop publishing.[22] Since the six-primary LED backlight ignites two spectrally different emissions sequentially, the following issues arise:

(1) luminance is low at $80 \, \text{cd}/\text{m}^2$ due to low light-emission duty of LEDs,
(2) flicker is observed due to the alternating emission.

For (1), the luminous efficacy of LEDs should be improved. Alternatively the number of LEDs used in the backlight unit (now 216) may be increased. For (2), more accurate signal processing is required.

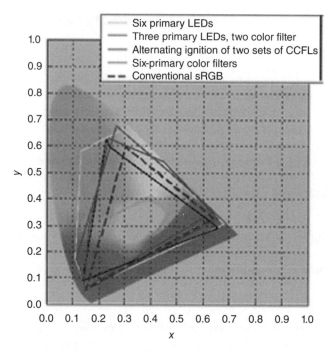

Figure 3.11   Color gamut comparison of various types.

Table 3.4   Color gamut comparison of various types (areas on CIE x, y chromaticity diagram; the area of sRGB = 100).

	Area on CIE x, y chromaticity diagram
Six-primary LED backlight[10]	175
Three-primary LED backlight with 2 color filters[11]	136
Alternating ignition of two sets of CCFLs[21]	115
Six-primary color filters[9]	138
Conventional sRGB	100

References

[1] IEC 61966-2-1 (1999) 'Multimedia systems and equipment – colour measurement and management – Part 2-1: Colour management – Default RGB colour space – sRGB', Oct. 18, 1999.

[2] IEC 61966-2-1 (2003) Amendment 1, 'Multimedia systems and equipment – colour measurement and management – Part 2-1: Colour management – Default RGB colour space – sRGB', Jan. 23, 2003.

[3] IEC 61966-2-2 (2003) 'Multimedia systems and equipment – colour measurement and management – Part 2-2: Colour management – Extended RGB colour space – scRGB', Jan. 23, 2003.

[4] JEITA CP-3451-1 (2003) 'Video file format standards for digital still cameras', Exif 2.21, Sep. 2003.

[5] IEC 61966-2-4 (2006) 'Multimedia systems and equipment – colour measurement and management – Part 2-4: Colour management – Extended – gamut YCC colour space for video applications – xvYCC'.

[6] IEC WD 61966-2-5 (2004) 'Multimedia systems and equipment – colour measurement and management – Part 2-5: Colour management optional RGB colour space – opRGB", Oct. 22, 2004.

[7] Shibuya, *et al.* (2003) ITE Convention, Inst. Image Information and Television Engineers.

[8] Sugiura, H. *et al.* (2004) *SID '04 Digest*, pp. 1230–1233.

[9] Yang, Y.-C. *et al.* (2005) *SID '05 Digest*, pp. 1210–1213.

[10] Sugiura, H. *et al.* (2005) *SID '05 Digest*, pp. 1124–1127.

[11] Roosendaal, S. J. *et al.* (2005) *SID '05 Digest*, pp. 1116–1119.

[12] DDWG (1999) *Digital Visual Interface revision 1.0*, Apr. 2, 1999.

[13] VESA (2004) *Display Data Channel Command Interface Standard, Version 1.1*, Oct. 29, 2004.

[14] VESA (2003) *VESA Monitor Control Command Set (MCCS) Standard, Version 2*, Oct. 17, 2003.

[15] Sugiura, H. *et al.* (2001) *Proc. SPIE*, **4300**, p. 278.

[16] Sugiura, H. *et al.* (2002) *Proc. SPIE*, **4657**, pp. 54–61.

[17] Sugiura, H. *et al.* (2002) *Proc. ICIS '02 Tokyo*, pp. 343–344.

[18] Sugiura, H. *et al.* (2002) *SID '02 Digest*, pp. 288–291.

[19] Sugiura, H. *et al.* (2002) *Proc. EURODISPLAY '02*, pp. 379–382.

[20] 'Munsell Color Cascade' data measured at the National Physical Laboratory, UK, and supplied by Dr Michael R. Pointer.

[21] Jak, M. J. J. *et al.* (2005) *SID '05 Digest*, pp. 1120–1123.

[22] Sugiura, H. *et al.* (2005) *Display Monthly*, **11**, 8–17.

4

Reduction of Backlight Power Consumption of LCD-TVs

T. Shiga

The University of Electro-Communications

4.1 Introduction

For large screen LC-TVs, the power consumption of backlight units is quite high and therefore its reduction is a priority. In general, the power reduction of the backlight unit is accomplished by improving the luminous efficacy of the light source and also by improving optical components such as reflectors and diffusers. In addition to these, a dimming backlight technique was recently introduced. The conventional backlight always emits light at a constant luminance. In contrast, the luminance for the dimming technique varies along with the image contents. This technique may be referred to as a dynamic backlight technique. Other examples of dynamic backlight techniques are blinking backlight, scanning backlight[1] and field sequential backlight.

LCD Backlights   Edited by Shunsuke Kobayashi, Shigeo Mikoshiba and Sungkyoo Lim
© 2009 John Wiley & Sons, Ltd.

4.2 Display Method of LCD and Power Reduction

For conventional LC-TVs the backlight unit is always operated at full luminance and full power regardless of the images to be displayed. In contrast,
for emissive displays such as CRTs, PDPs and OLEDs, the power depends
on an average picture level (APL) of the TV contents. The APL for an ordinary broadcast TV signal is approximately 25%, implying that the backlight
power could be reduced to about 25%.

One of the methods of reducing backlight power is to modify the gamma
curve and reduce the backlight luminance according to the histogram of
input image as explained in Figure 4.1. This method was originally intended
to obtain high dark room contrast ratios.[2,3] With this method, the light
leakage from the LC panel is reduced and the dark part of the image becomes
darker, resulting in greater contrast.[4,5]

The other method of reducing the backlight power is to employ an adaptive dimming technique.[6] The basic principle of the technique is as follows.
The peak value of the input signal across the entire display area is detected
and amplified to the 255th level (for an 8-bit expression) by a factor $1/k$. At
the same time the lamp luminance is dimmed by a factor k so that the luminance of the original image is retrieved, while the lamp power is reduced
by a factor k.

        (a)   (b)   (c)

Figure 4.1 Power saving with modification of gamma curve and backlight
dimming: (a) original image, (b) image with reduced gamma without
dimming, (c) image with reduced gamma with dimming.

4.3 Principle of the Adaptive Dimming Technique

A basic principle of power saving with adaptive dimming is explained in Figure 4.2. It is assumed here that the transmission of the liquid crystal panel is proportional to the input signal. When the input TV signal level is 25%, for instance, in the conventional technique, the transmission of the liquid crystal is reduced to 25% while keeping the backlight luminance unchanged. Power consumption of the backlight unit is constant, independent of the TV signal. With the adaptive dimming technique, the 25% TV signal level is expanded to 100%, whereas the backlight is dimmed down to 25%. The power consumption of the backlight, therefore, is reduced to 25%, while the original picture can be reproduced.

An example of a CCFL backlight unit with dimming capability is shown in Figure 4.3. Here x and y denote the horizontal and vertical position of the backlight unit. It is assumed that the luminance distribution along the x direction is uniform. Each lamp is equipped with an inverter with a dimming capability. The dimming factor (which is the ratio of the luminance after dimming to the luminance before dimming) is defined as k ($0 \leq k \leq 1$). Figure 4.4 shows luminance distributions of the respective eight lamps, $L_m(y)$, $m = 1$–8, and also a distribution $L_b(y)$ when all the lamps are fully ignited ($k = 1$). Here $L_b(y) = \sum_{m=1}^{8} L_m(y)$. The luminance of the original image, $B(x, y)$, for the original signal, $s_i(x, y)$, is obtained from

$$B(x, y) = L_{b\min} T(s_i(x, y))$$

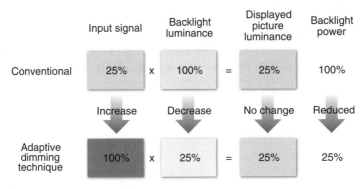

Figure 4.2 Principle of power saving with the adaptive dimming technique.

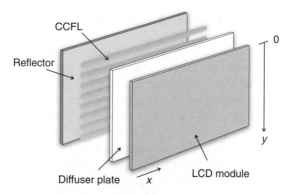

Figure 4.3 Eight CCFL backlight unit.

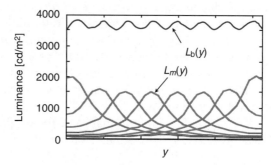

Figure 4.4 Luminance distribution after diffuser plate. $L_m(y)$: luminance of lamp m; $L_b(y)$: luminance when all the lamps are ignited.

where L_{bmin} is the minimum luminance of $L_b(y)$ and $T(s)$ is the transmission characteristic of the liquid crystal for signal s ($0 \leq s \leq 255$). $B(x, y)$ is the target luminance after application of the dimming technique.

There are several methods for determining the set of dimming factors, [k]. One example is shown in the flowchart of Figure 4.5. First, a set of k ([k]) is arbitrarily chosen. Then the backlight luminance after the adaptive dimming, $L_a(y)$, is calculated. If the following equation is satisfied, then the backlight power can be calculated:

$$L_a(y)T(255) \geq B_{max}(y).$$

Here $B_{max}(y)$ is the peak luminance of $B(y)$. The left-hand side of the equation represents the maximum luminance for $L_a(y)$. If $L_a(y)T(255)$ is smaller than

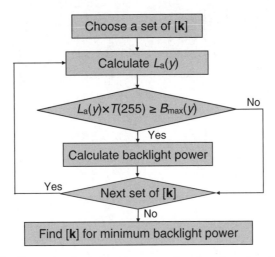

Figure 4.5   Flow chart of signal processing for dimming factors (**k**).

$B_{max}(y)$, $B(x, y)$ cannot be reproduced. After the calculation of power, the next set of k is chosen. This loop is repeated until a set of k which gives minimum backlight power is found. The input signal after the adaptive dimming, $s_a(x, y)$, is obtained from the equation:

$$L_a(y)T(s_i(x, y)) = B(x, y).$$

In this way, $L_a(y)$ and $s_a(x, y)$ are determined so that the output luminance before the dimming can be reproduced.

Figure 4.6 shows simulated results. The APL of the signal is 25%. With the adaptive dimming technique, backlight luminance $L_b(y)$ is reduced to $L_a(y)$ and the signal $s_i(y)$ is increased to $s_a(y)$. The backlight power consumption with the dimming is reduced to 34% of the original value.

4.4 Adaptive Dimming Control and Power Consumption

4.4.1 A Use of Optical Isolator

It can be seen from Figure 4.4 that the output of each lamp spreads to an appreciable vertical distance. Due to the spread of the lamp output, the peak luminance of $L_m(y)$ is only 40% of $L_b(y)$. With this luminance distribution, the adaptive dimming technique is not effective for a picture having a bright

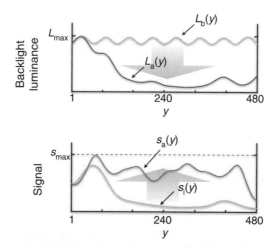

Figure 4.6 Simulated results of adaptive dimming: (a) luminance distribution before adaptive dimming, $L_b(y)$, and after adaptive dimming, $L_a(y)$; (b) original input signal, $s_i(y)$, and signal after adaptive dimming, $s_a(y)$.

spot on a dark background as in the picture 'sunrise' of Figure 4.7(a). For the picture entitled 'sunrise', all of the eight lamps should be almost fully turned on in order to obtain the high luminance of the sun. One of the methods to overcome the problem is to optically isolate the lamps from one another by providing reflecting optical isolators, suppressing the spread of the lamp light and also increasing the peak luminance of $L_m(y)$.[7]

Table 4.1 lists the experimental results of power reduction with or without optical isolators for the pictures shown in Figure 4.7.[7] The APL of both pictures is 28%. For 'sunrise' the power was reduced significantly, but there was only a limited improvement for the picture 'garden'. It can be concluded that the optical isolators are especially useful for pictures with a low APL but having bright spots.

4.4.2 Comparison of 0D, 1D and 2D Dimming[8]

Figure 4.8 shows zero-, one- and two-dimensional (0D, 1D and 2D) backlight unit configurations. The conventional backlight without individually controllable inverters can use only the 0D dimming (uniform dimming). 1D dimming (line dimming) is applicable to CCFLs, HCFLs and EEFLs, and 2D dimming (local dimming) is applicable to LED backlights as well as matrix type FED, OLED and plasma backlights.

(a) Sunrise

(b) Garden

Figure 4.7 Sample pictures: (a) sunrise and (b) garden.

The 0D, 1D and 2D dimming controls are examined with an LED backlight having optical isolators as shown in Figure 4.9.[8] The dimming experiments were performed using a 10-s movie 'sunset'. The movie has a very bright spot on a dark background. Figure 4.10 shows measured variations of power consumption with respect to time for the 0D, 1D and 2D dimming controls.

Table 4.1 Power saving with and without optical isolators.

Test content	Sunrise	Garden
APL	28%	28%
without adaptive dimming	100%	100%
with adaptive dimming without optical isolators	83%	56%
with adaptive dimming with optical isolators	53%	54%

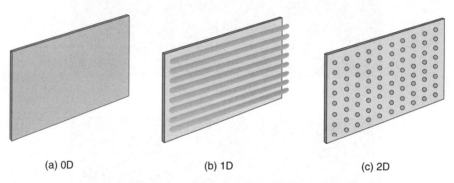

(a) 0D   (b) 1D   (c) 2D

Figure 4.8 Dimensional dimming of backlights.

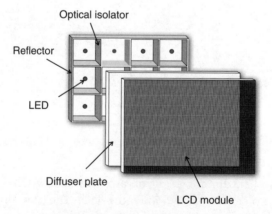

Figure 4.9 LED backlight unit.

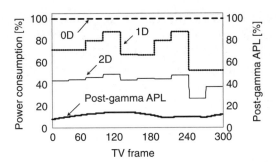

Figure 4.10 Variation of power consumption for 0D, 1D and 2D dimming controls.

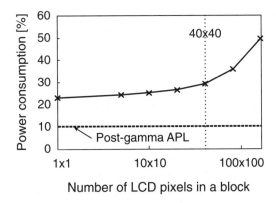

Figure 4.11 Block size of backlight unit versus power consumption.

The refresh rate of the dimming factor was 1 s. Variation of the post-gamma APL is also plotted in Figure 4.10. Since there always is the very bright spot in the picture, the backlight luminance cannot be dimmed for the case of 0D dimming. For the case of 1D and 2D dimming, the averaged power was 72% and 43% of the original value, respectively. The results show that the 2D control is the best.

4.4.3 Choice of Block Size

The LED backlight unit shown in Figure 4.9 is divided into 3×4 blocks. The effectiveness of power reduction depends on the number of corresponding LCD pixels in a block. Figure 4.11 shows calculated variations of power consumption with respect to the number of corresponding LCD pixels in a block.[8] The sample movie 'sunset' was used for the simulation. A reduction in the number of pixels in a block, that is, an increase in the total number of blocks in the backlight unit, reduces the power consumption, although no significant power reduction is observed as long as the number of the pixels in a block is kept below 40×40.

4.4.4 Independent RGB LED Control

For LED backlights, independent luminance control of R, G and B are possible, unlike the white-emitting CCFL backlights. Consider two extreme cases. One is to reproduce a white-balanced picture. In this instance there is no difference in the power reduction for both LEDs and CCFLs. The other extreme is to reproduce a pure red image at its peak luminance. For CCFLs, the dimming factor is 1, that is no dimming, and hence no power saving. For LEDs, on the other hand, the dimming factors for R, G and B are, respectively, 1, 0 and 0. As a result the power consumption is reduced to a third. The power saving depends largely on the picture to be displayed.

4.5 Other Features of the Adaptive Dimming Technique

The adaptive dimming technique improves the contrast ratio since the leakage light from the backlight is reduced when the light sources are dimmed down. Also a non-uniform luminance distribution of the backlight unit, caused by the structural variations or by the optical isolators, can be compensated for by the dimming technique introduced here.

For the incorporation of 2D dimming, the use of an LED backlight is the best choice at present. LED backlights have features including fast response, wide color gamut and they are mercury-free. On the other hand, LED backlights have negative features such as high power consumption, differential aging characteristics between the LED components and temperature dependence of color and luminance. Use of the adaptive dimming technique for an LED backlight can solve some of these problems.

Since the signal is magnified almost to the maximum level, the gray scale capability is enhanced especially for low luminance level pictures. Since the human eye has a logarithmic sensitivity with respect to luminance, precise gray scale expression is required especially for low luminance levels. Eleven-bit-equivalent resolution of gray scale has been experimentally proven with the adaptive dimming technique for low luminance levels.[9]

The technique can be used together with field sequential color display, eliminating color filters. Also the blinking/scanning backlight technique can readily be incorporated in order to improve the picture quality of moving images.

References

[1] Fisekovic, N. *et al.* (2001) *Proc. Asia Display/IDW '01*, pp. 1637–1640.
[2] Funamoto, T. *et al.* (2000) *Proc. IDW '00*, pp. 1157–1158.
[3] Choi, I. *et al.* (2002) *ISLPED '02*, pp. 112–117.
[4] Cheng, W.-C. *et al.* (2004) *IEEE Trans. on Consumer Electronics*, **50**, 25–32.
[5] Iranli, *et al.* (2005) *Proc. Design Automation and Test in Europe*, pp. 346–351.
[6] Shiga, T. and Mikoshiba, S. (2003) *SID '03 Digest*, pp. 1365–1367.
[7] Shiga, T. *et al.* (2005) *SID '05 Digest*, pp. 992–995.
[8] Shirai, T. *et al.* (2006) *SID '06 Digest*, pp. 1520–1523.
[9] Shimizukawa, S. *et al.* (2006) *Proc. IDW '06*, pp. 1743–1746.

5

Notebook PC/Monitor Backlights

B. H. Hong

Kwangwoon University

5.1 Introduction

LCDs are non-emissive type displays, and for this reason they are inferior to emissive type displays in terms of legibility. Nevertheless, LCDs are being utilized in a wide variety of applications ranging from small size mobile displays to large size TVs. Even 100-inch diagonals will be on the market. Historically the most important factor which caused the present widespread usage was the popularization of computers for personal use which started in the late 1980s. In addition, LCDs are replacing the traditional CRTs for desktop computer monitors. Reflecting this situation, significant advances in terms of performance and cost reduction of LCDs are being achieved. Among these advances, innovations in backlight units are playing a major role in LCD technology, since the image quality of LCDs is governed by the backlights. Advancements in general lighting technology enabled CCFLs to be used as backlights which overcame the shortcomings of LCDs. Thus, it is necessary to develop high-efficiency lamps, optical systems and driving circuits to provide high-quality backlight units.

LCD Backlights   Edited by Shunsuke Kobayashi, Shigeo Mikoshiba and Sungkyoo Lim
© 2009 John Wiley & Sons, Ltd.

5.2 Characteristics Required for Backlights

Characteristics required for notebook PC and monitor backlight units will be discussed. Firstly, luminance of the backlight unit with a single lamp must be at least $1900\,cd/m^2$. Several examples of backlights will be introduced. One of the examples features a high luminance of $2700\,cd/m^2$ using a single lamp with lens films. Another is a low cost backlight unit without any reflective polarizer films. There also is one which can achieve $3600\,cd/m^2$ using two lamps that is designed for enjoying movies with DVDs. Further, a novel backlight for PC monitor achieves $5000\,cd/m^2$ using four or six lamps.

Luminance uniformity of the backlight unit is defined by the ratio of luminance at the peripheral area to luminance at the central area. There are two definitions for the uniformity. Draw on the screen five horizontal and five vertical lines to make 25 cross points. The first definition, $UR(1)$, is expressed as:

$$UR(1) = \frac{[Minimum\ luminance\ of\ the\ 25\ cross\ points]}{[Luminance\ at\ the\ center]}$$

and the second definition, $UR(2)$, is expressed as:

$$UR(2) = \frac{[Minimum\ luminance\ of\ the\ inner\ 9\ cross\ points]}{[Luminance\ at\ the\ center]}.$$

For note PC backlights, $UR(1)$ and $UR(2)$ must be over 65% and 80%, respectively; for monitors, $UR(1)$ and $UR(2)$ must be over 75% and 90%, respectively. Although the angular dependence of the luminous uniformity is not clearly defined, no stained spots are desired. It is recommended that color is expressed to the accuracy of $1/1000$ in the CIE 1931 (x, y) color space. For light sources of monitor backlights, CCFLs are widely used. It is preferred that the lamps should not generate much heat so that they do not heat nearby objects including the LCD panels. Degradation of image quality caused by heat should be avoided.

5.3 Optical Systems for Backlights

The optical system of an LCD is made up of the following parts: the first polarizer, a liquid crystal cell, color filters and the second polarizer (called an analyzer). Light emitted from a lamp transmits through these optical components in the above sequence. RGB color filters are installed at every dot element. Thus, only a portion of polarized light and the optical spectrum contributes to the output. A backlight unit consists of a light source, a reflector, a light-guide plate used to convert light from a point source or a line source to an area source, a diffuser for obtaining uniform luminance and a lens film (prism sheet) or a reflective polarizer film for luminance enhancement.

Table 5.1 lists optical components used for backlight units for note PCs, monitors and LC-TVs. For the first two categories a light source is located at the side of a light-guide plate and, for a TV, multiple light sources are located behind the LCD. A large number of light sources are needed to obtain high luminance. For note PCs and monitors, a sophisticated optical system is needed for obtaining uniform luminance, but the number of light sources is less than that for TV use.

Recent technologies have achieved progress in areas such as cost reduction, low power consumption, thickness reduction and weight reduction. Further cost reduction, especially of optical sheets, and a use of fewer components may bring a larger scale and wider penetration of LCDs into the market.

Table 5.1 Optical components used in backlight in units.

Function	Note PC	Monitor	TV
light source	CCFL, LED	CCFL, LED	CCFL, LED
arrangement of light sources	light guide plate	light guide plate	direct lit
optical coupling of lamps	lamp reflection sheet	lamp reflection metal	none
light guide	light diffusion dot pattern	light diffusion dot pattern	none
homogenization of luminance	light diffusion sheet	light diffusion sheet	light diffusion plate/sheet
directionality	two prism sheets	one prism sheet	one prism sheet
optical	none	polarization reflection film	polarization reflection film
polarization	CCFL, LED	CCFL, LED	CCFL, LED
protection film	none	with protection film	none

5.4 Light Sources for Backlights

Cold cathode fluorescent lamps (CCFLs) are predominantly used as light sources for backlights. Generally, an Ne–Ar gas mixture is admitted into the lamp together with a small amount of Hg. The mixture ratio, pressure and tube diameter depend on the purpose for which it is to be used. An addition of Ar between 3% and 10% results in a lowering of the ignition voltage due to the Penning effect,[1] and also suppresses red emission from Ne during the early stage of ignition when Hg pressure is not sufficiently high. During the discharge, the amount of Hg decreases due to the formation of an amalgam when Ar ions collide with the cathode. For monitors, the Ar content has to be reduced but for note PCs the content has to be increased, since the luminance decreases as the Ar content increases when the electrical current exceeds 6 mA.

In the low current region, higher luminance is obtained with higher gas pressures, and in the high current region, higher luminance is obtained with lower gas pressures. On the other hand, in the low current region, the ignition voltage becomes lower as the gas pressure is reduced, and in the high current region, the ignition voltage does not depend on the gas pressure. Normally, a total pressure of 60–80 Torr is chosen. The optimum Hg pressure is higher with a smaller tube diameter, resulting in higher luminance due to higher collision probability between electrons and Hg atoms. The total output flux, however, decreases with a smaller tube diameter. Appropriate diameters would be 2.6 mm for monitors and 2.0 mm for note PCs.

The life of CCFLs is defined by the time the lamp luminance is reduced to 50% of the initial value. Degradation of the luminance occurs due to the following factors:

(1) vacuum ultraviolet radiation of 185 nm emitted from Hg damages the phosphor;
(2) adhesion of Hg particles onto the phosphor surface blocks VUV and visible light;
(3) Na atoms coming out of the tube glass absorb visible light.

By forming a protecting layer between the phosphor and tube and/or between the phosphor particles, the life of CCFLs may be extended to 70 000 hours compared with the conventional 50 000 hours.[2] In order to broaden the color gamut, new phosphors are being actively investigated. If a CCFL is driven with an asymmetrical voltage waveform, then non-uniform Hg distribution is created, which is an undesirable condition. It is still necessary to conduct research and developments on electrode materials, lamp structures, gas content and gas pressure to obtain higher luminance and longer life, since a high discharge current operation for obtaining a high luminance reduces the life.

5.5 Optical Components of Backlights

Figure 5.1 illustrates how the total luminous flux and luminance vary when passing through respective optical components of the backlight unit which utilizes an edge type light-guide plate. In the diagram, the line with circular dots indicates the total luminous flux that was measured with an integrating sphere. The line with square dots is the luminance of the incident light into an LCD panel, measured at normal to the panel. The luminous flux decreases at each optical component, but the luminance increases.

5.5.1 Lamp Reflector

It can be understood from Figure 5.1 that the largest loss occurs at the lamp reflector, where the flux is reduced by 25%. Only 75% of the initial flux can enter the light-guide plate (LGP). Figure 5.2 shows examples of lamp reflectors for note PCs and monitors. The drawings show that a large portion of

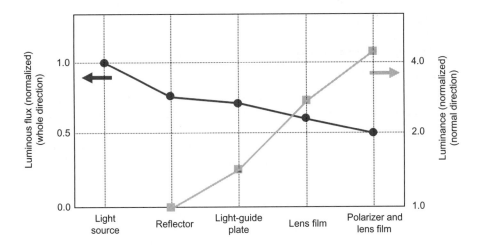

Figure 5.1 Variations of total luminous flux and luminance.

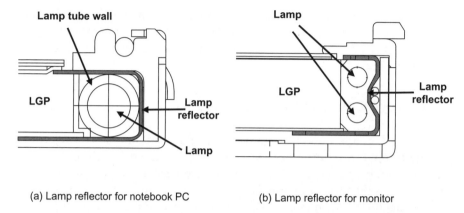

(a) Lamp reflector for notebook PC　　　　　(b) Lamp reflector for monitor

Figure 5.2　Lamp reflectors for (a) note PCs and (b) monitors.

the light emitted away from the lamp cannot enter the light-guide plate even with the lamp reflectors.

There are two kinds of lamp reflectors: one is a metal type and the other is a sheet type. The metal type is normally used for monitors and made of SUS or aluminum. The metal is either coated with silver or a reflective sheet is glued onto it. The sheet type is used for notes and made of PET on which a reflective sheet is glued. The sheet type is less expensive but provides less reflection.

5.5.2 Light-guide Plate

A backlight unit needs a light-guide plate which expands the light from a CCFL to match the area of the display. The most common material for the light-guide plate is PMMA (polymethylmethacrylate). This material is an inexpensive, non-fragile, solid and has an excellent high optical transparency, but is susceptible to heat and water attack. The relative density of PMMA is 1.19. An alternative material is Zenor, an olefin resin. This is used for its low weight with a relative density of 1.019, but it is soft and difficult to make with an injection molding.

The light-guide plate is classified into two categories: one is a flat type and the other is a wedge type (Figure 5.3). The flat type is used for monitors which require high luminance. The high luminance can be obtained by arranging lamps on both sides or on four sides of the light-guide plate. The wedge type is commonly used for portable note PCs whose power consump-

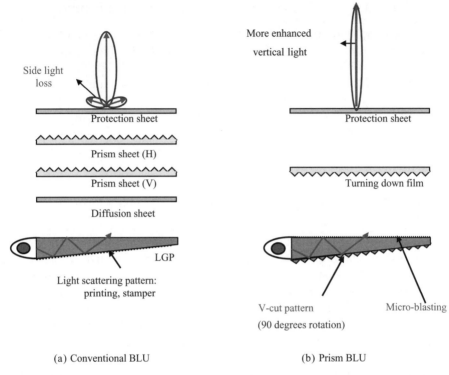

Side light
loss

More enhanced
vertical light

Protection sheet

Prism sheet (H)

Prism sheet (V)

Diffusion sheet

LGP

Light scattering pattern:
printing, stamper

Protection sheet

Turning down film

V-cut pattern
(90 degrees rotation)

Micro-blasting

(a) Conventional BLU   (b) Prism BLU

Figure 5.3 (a) Conventional backlight unit, and (b) prism backlight unit, both for note PCs.

tion should be low. A lamp is placed at one side of a light-guide plate. The flat type light-guide plates are fabricated by cutting a large acrylate plate which is prepared by a push molding, while the wedge type is fabricated by an injection molding. Features of the wedge type are that it has a high optical utilization efficiency and it is easy to fabricate as a thin plate. Light incident on the light-guide plate cannot emerge from the plate due to the total internal reflection. In order to redirect and extract the light, a scattering pattern is provided at the bottom surface of the plate. There are two ways for forming the pattern: one is with printing and the other is without printing. For the printing method, light scattering ink is screen printed at the back of the plate. The non-printing method uses injection molding to form a surface having a special shape. Backlight units with the diagonal exceeding 17-inch commonly use the printing method.

5.5.2.1 Light-guide Plate Fabricated with Printing Method

When forming a pattern with the screen printing method, SiO_2 or TiO_2 pastes are commonly used. SiO_2 has a refractive index of 1.6 which is almost the same as that of PMMA, while that of TiO_2 is around 2.8. The SiO_2 particles scatter the light, which is then reflected by a reflective sheet attached to the back of the light-guide plate. The TiO_2 pattern scatters and reflects the light. Recently, SiO_2 has been favorably utilized for its low scattering loss. The shape of the printed pattern is circular, ellipsoidal, square or hexagonal. The circular and ellipsoidal patterns are easy to print; but the square and hexagonal patterns are easy to design.

5.5.2.2 Light-guide Plate Fabricated with Non-printing Method

There are several injection molding methods for fabricating the light-guide plates: metal shaping, pattern formation and others, which are selected depending on the shape of the pattern. Metal shaping is done by engraving a pattern directly onto the metal mold core. Alternatively a pattern is made on the metal mold core by using a stamper. The pattern of the stamper is formed by chemical etching, sandblasting, Ni electro-plating or screen printing. For chemical etching, a photoresist material is coated on the back surface of the metal stamper, and a patterned film is placed on it. Then exposure, development and etching are carried out. Although the fabrication is simple, it is not so easy to obtain good reproducibility of the pattern size, depth and optical uniformity due to difficulties in temperature control of the etchant and also control of the etching time. The optical efficiency of the light-guide plate fabricated by this method is almost the same as that fabricated by the printing. Sandblasting ensures reproducibility of the patterns with high optical uniformity. For the etching method, the etched profile tends to become a hemisphere and this results in the production of a high optical throughput. Generally the pattern for screen printing is hemispherical. Formation of a square pattern at the back of the light-guide plate is also possible.

Another method is to form a pattern on the upper surface of a light-guide plate with an injection mold. In this method, it is possible to control the distribution of the light output by controlling the density of fine reticles formed on the surface of the metal mold. This results in an increase in the luminance of 10% or more. There is another method which increases the optical efficiency by 50%. The method is to form a sheet having an inverted prism, with the prism oriented downward, as will be explained below.

5.5.2.3 Prism light-guide Plate for Note PCs

With an increase of LCD resolution, the aperture ratio of each pixel decreases, requiring higher luminance of the backlight unit. For this purpose, new light-guide plates are being developed aiming at not only efficient diffusion but also providing directionality of the output light. The introduction of multiplefunctions to the light-guide plate also results in lower power consumption as well as reduced thickness and weight of the backlight unit.

The use of a prism pattern[3] features high luminance, low power consumption and low weight even with a thickness of a quarter of the conventional type. The essence is that the light-guide plate is integrated with the other optical components shown in Table 5.1. Figure 5.3 compares a conventional backlight unit and a prism backlight unit for note PCs. The conventional backlight unit of Figure 5.3(a) uses a light-guide plate, a diffusion sheet, two lens sheets (or prism sheets) and a protection sheet. The prism backlight unit of Figure 5.3(b), on the other hand, uses a light-guide plate, an inverted prism sheet and a protection sheet but not a reflective polarizer. This brings an increase of luminance and a reduction in the cost. The light-guide plate is installed with a V-cut pattern at the back of the plate and with a matt treatment at the front of the plate, hence sometimes referred to as a matt light-guide plate. Figure 5.4 explains the processes for making the V-cut

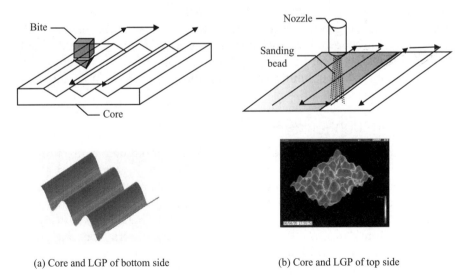

(a) Core and LGP of bottom side   (b) Core and LGP of top side

Figure 5.4 Formation of core and surface structure of a light-guide plate.

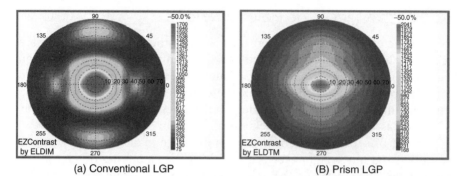

(a) Conventional LGP   (B) Prism LGP

Figure 5.5 Emission angles of (a) conventional and (b) prism backlight units.

patterns and the matt treatment on the metal core. Photos of back and front surfaces of the light-guide plate are also shown.

Figure 5.5 compares the emission angle dependencies of the conventional and prism backlight units. Although the conventional backlight unit emits light in the vertical direction as well as in non-vertical directions, the prism backlight unit emits most of the light only in the vertical direction. The light emitted at 70° off-normal angle for the conventional backlight unit, for instance, cannot enter the LCD panel and is therefore wasted. The degree of convergence of light into the vertical direction strongly depends on the shape and size of the matt pattern of Figure 5.4(b). Although the prism backlight unit has various features, only a few manufactures are capable of fabricating it, limiting the penetration of this technology into note PCs.

5.5.2.4 Prism Light-guide Plate for Monitors

Prism light-guide plates fabricated by an injection molding are also utilized for monitor backlights. In this case, a prism pattern is also formed at the front surface. The prism patterns at the bottom of the plate are made parallel to the lamp, giving uniform luminance distribution. The prism patterns at the front of the plate collect light from the normal direction, producing an increase of luminance of 10%. Injection molding is useful for fabricating light-guide plates up to 15 inches, but not for larger sizes, since the molding requires a long cycle time which reduces productivity.

References

[1] Razer, Y. P. (1987) *Gas Discharge Physics* Springer-Verlag, p. 59.
[2] Matsuo, K. (2003) 'Novel Technologies of Backlight Phosphors and Protection Layers', *Display Monthly, Technotimes*, **9**.
[3] Okada, H. (2004) 'Development and Trend in the Prism Sheet for Backlight', *Display Monthly, Technotimes*, **10**, p. 004.

6

Backlights for Handheld Data Terminals

S. Aoyama

Omron Corporation

6.1 Introduction

In 2007, shipments of mobile phones reached some 930 million units.[1] One of the major driving forces behind this growth is that cell phones introduced color liquid crystal displays in 2000. Before that, the mobile phones had been used just for voice communication. Their role has since developed to include communication via text and still images and, thanks to the incorporation of digital communication technology, these devices have evolved into comprehensive data communication tools that can also handle video images. The growing popularity of video content has led to a demand for liquid crystal displays that are not only slimmer and consume less power, but also have higher image quality and resolution. LED backlights play an important role in supporting this trend toward video content. Consequently, LED backlights are subject to the same demands as liquid crystal displays with regard to improved portability and image quality.

The basic structure and operating principles of LED backlights are described in this chapter. Also how the above mentioned demands for brighter, slimmer and lighter LED backlights are being addressed is

LCD Backlights   Edited by Shunsuke Kobayashi, Shigeo Mikoshiba and Sungkyoo Lim
© 2009 John Wiley & Sons, Ltd.

discussed considering the latest technical trends in their constituent parts such as light guides and diffuser sheets.

6.2 Basic Structure and Principles of LED Backlights

6.2.1 Structure of a Backlight with Two Prism Sheets

Figure 6.1 shows the basic structure of the most common type of LED backlight, which incorporates two prism sheets. The basic constituent parts of the LED backlight are the LED module light source, a light guide, a reflector sheet, a diffuser sheet and two prism sheets. The brightness of the LED backlight largely depends on the light-emitting efficiency of the LED module. Back in 2000 when color displays first appeared, the light-emitting efficiency of LED backlights was about 10 lm/W. However, their efficiency has been increasing every year and it is currently about 70 lm/W. By 2010, it may be possible to achieve 100 lm/W, which is about twice the luminous efficiency of the cold-cathode fluorescent devices. Together with the latest RoHS directives which impose broader restrictions on the use of mercury, and the

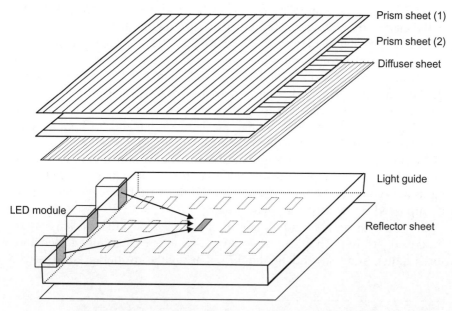

Prism sheet (1)

Prism sheet (2)

Diffuser sheet

Light guide

LED module

Reflector sheet

Figure 6.1 Basic structure of an LED backlight unit.

continuing industrial trend towards energy-saving technology, LEDs are likely to be used increasingly in medium-size backlights for PDAs and car navigation systems.

The light guide serves to spread the light of the point sources of the LED modules across the plane of the backlight. An arrangement of diffuser dots and/or prisms is formed on the back surface of the light guide so as to achieve an efficient and uniform radiation of light from the light-guide surface. Recent advances in micromachining technology have made it possible to achieve greater details with higher precision, resulting in substantial improvements in emission efficiency and uniformity. The thickness of the light guide was about 0.7 mm for the initial color devices, but with modern technology this has been reduced to 0.4 mm in the latest products. Most of the light guides are made from materials such as acrylate and polycarbonate resins which have superior optical and mechanical properties, but cycloolefin polymer resins are also used in cases where low weight is required. The prism sheets are still essential parts of the LED backlight unit which perform multiple functions of extracting light from the light guide and condensing it into a suitable angular range, while returning the components of light at other angles to the light guide to increase its uniformity, and reducing visibility of defects in the light guide. A backlight normally uses two prism sheets in which the prisms are arranged at right angles. Compared with a single prism sheet, the use of two prism sheets results in a better condensing performance, better luminance uniformity and reduced defect visibility.

The reflector sheet is made of a silver foil formed on a sheet, or of a white sheet of PET. The diffuser sheet can be adjusted to a required level of diffusion by selecting a suitable degree of haze. The properties of the reflector sheet and diffuser sheet – such as their reflectivity, reflection characteristics and diffusion characteristics – are selected to ensure that the radiated intensity and uniformity of light emitted from the prism sheet satisfy the desired properties in combination with the light emission characteristics of the light guide.

There are two ways to increase the brightness of the backlight unit based on the construction of Figure 6.1. One is to increase the coupling efficiency of the materials through which the light propagates from the LED module so that it can be radiated from the prism sheet without loss, and the other is to constrict the intensity distribution of the light emitted from the prism sheets so that a higher intensity is radiated in the direction normal to the plane of the backlight. However, since the structure of the prism sheet is designed to maximize the radiated intensity, there is generally little freedom in the design. As a result, we are restricted to making more efficient use of the light emitted from the LEDs by increasing the coupling efficiency of the materials. With current technology, it is possible to achieve a coupling efficiency of 95% between the LED module and light guide, a confinement efficiency of 95% in the light guide (whose efficiency increases by reducing the leakage of light from the end surfaces of the light guide) and an efficiency of about 85% in the reflector sheet. As a result, about 75% of the light from the LED module is emitted from the prism sheet. This indicates that even in an ideal case it is only possible to achieve a brightness increase of 25%.

On the other hand, thinner backlights are implemented by reducing the thicknesses of the light guide and other constituent sheets. Approximate thicknesses of the reflector sheet, diffuser sheet and prism sheets were 80 μm, 60 μm and 60 μm, respectively. These have been reduced to about a half due to the above-mentioned trend towards reduced light-guide thickness. Considering the pattern-forming properties and mechanical strength of the light guide and prism sheets, a major breakthrough is needed to achieve further reductions of thickness.

6.2.2 Principles of LED Backlights

An LED backlight unit has three functions: (1) converting the light flux from one or more LED point sources into a planar light source, (2) emitting the light flux from the upper surface of the light guide, and (3) condensing the emitted light with the required level of directivity. These functions are further explained below.

Figure 6.2 illustrates the behavior of a link of an LED module light source with the light guide. The LED module is directly abutted against the edge surface of the light guide. The light radiated from the module is generally assumed to have a Lambertian intensity distribution. Consider the behavior of light rays in the y–z plane of Figure 6.2(a). Of the light rays emitted from the LED module, the rays parallel with the end surface of the light guide (traveling along the z-axis) are refracted at an angle α that satisfies the relationship $n_w \sin(\alpha) = 1$, where n_w is the refractive index of the light guide. These rays are incident on the upper surface of the light guide at an angle $\beta = \pi/2 - \alpha$. To avoid light radiated from the LED module being directly emitted from the upper or lower surfaces of the light guide, it is necessary to satisfy the relationship $n_w \sin(\beta) > 1$. A light guide that satisfies this condition must have a refractive index in the range $n_w > 1.4$.

Next, consider the behavior of light rays in the x–y plane of Figure 6.2(b). Here, it is necessary to minimize the so-called 'eyeball effect' of intensity

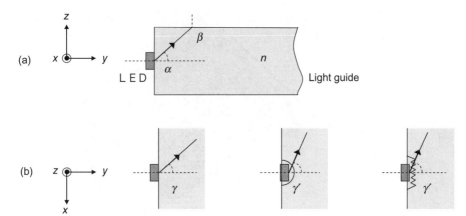

Figure 6.2 Behavior of light rays at the junction between the light-guide end and LED module.

variations originating from the point light sources. When the end surface of the light guide is flat, the incident angle of light from the end surface of the light guide is equal to the angle α mentioned above. If the light guide has a refractive index of 1.5, then α is 42°. Normally, multiple LED modules are provided at fixed intervals in parallel with the end surface of the light guide. If three LED modules are used in a light guide with a diagonal measurement of 2 inches, then the gap between these modules is 8 mm. Under this condition, a diffusion distance of at least 5 mm is needed for the light rays from the LED module to become superposed inside the light guide. The distance needs to be made as small as possible to reduce the frame thickness. One way of achieving this is to form an indented shape in the x–y plane in the vicinity of the LED module. With this configuration, however, it is difficult to achieve close coupling with the end surface of the light guide, which can lead to leakage of light rays in the z-direction and reduced efficiency of coupling between the LED module and the end surface of the light guide. A similar effect can be achieved by employing a structure where the end surface is textured by notching or sand-blasting in the z-axis direction. In this case, the light incident on the light guide is diffused by the texturing, allowing the design to employ a shorter diffusion distance.

Figure 6.3 shows an example of a pattern formed on the back surface of the light guide. In this example, LED modules are arranged along the left-hand edge. The pattern has the function of generating uniform emission

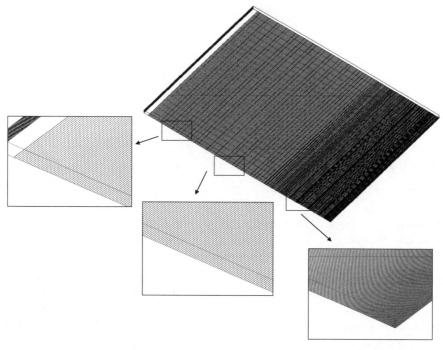

Figure 6.3 Diffusion pattern formed on the back surface of the light guide.

intensity by appropriately diffusing the light rays in the light guide. The diffusion pattern is required to be imperceptibly small when viewed from the upper surface of the light guide, and also should result in sufficient light output coming from the upper surface in order to obtain the desired level of brightness. Since the light guide contains more light rays in the vicinity of the LED modules, the pattern is designed to have lower density in this region, and the density increases with increasing separation from the LED module.

The prism sheets are provided for the purpose of condensing the light rays emitted from the light guide. As shown in Figure 6.4, the behavior of light rays in the prism sheet can be classified into three modes: an emission mode in which light is emitted from the upper surface, a recursion mode in which light is reflected back into the light guide and a high-angle mode in which light is emitted outside of the effective field of view. The emission mode corresponds to rays for which the half-width at half-maximum of the emitted intensity distribution is 20° or less.

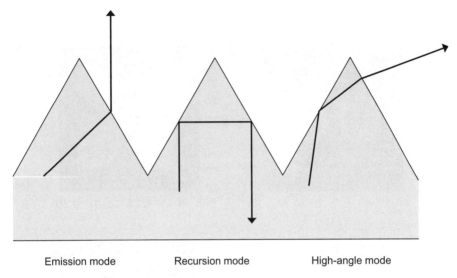

Emission mode   Recursion mode   High-angle mode

Figure 6.4 Behavior of light rays in the prism sheet.

Figures 6.5(a) and (b) illustrate the angle-of-incidence distribution of light rays onto the prism sheet. The vertical and horizontal coordinates express an incident angle (in degrees). The light rays incident on the prism sheet are assumed to have a Lambertian distribution. Figure 6.5(a) shows the result obtained with a single prism sheet whose ridge lines are arranged in the y-direction. The incident angle distribution is split symmetrically with respect to the prism ridge lines and is situated at approximately ±30° from the z-axis. It can also be seen that the distribution spreads over 41° in the x-direction and 50° in the y-direction. Of the total amount of light incident on the prism sheet, the amount of light emitted in the emission mode constitutes about 40%. When the prism ridge lines are arranged in the x-direction, the incident light distribution is split four ways at about 49° from the z-axis and spreads to an extent of 45° in both x- and y-directions as shown in Figure 6.5(b). In this case, the ratio of emission mode to the total input is 20%.

Figures 6.5(c) and (d) show the distributions of emitted light with one and two prism sheets, respectively. It can be seen that a stronger condensing effect is achieved with two prism sheets due to the higher selectivity of the prisms in transferring the incident angle distribution to the emission mode. The increased angular selectivity is also accompanied by an increased recursion mode, resulting in higher brightness uniformity.

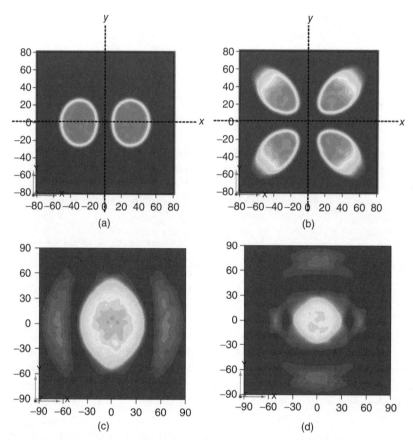

Figure 6.5 Distribution of the angles of incidence and emission of light rays at the prism sheet. (a) Distribution of the angle of incidence with a single prism sheet having ridge lines arranged in the *y*-direction. (b) Distribution of the angle of incidence with a single prism sheet having ridge lines arranged perpendicular to the *y*-direction. (c) Emission angle distribution with a single prism sheet. (d) Emission angle distribution with two prism sheets.

6.3 Constituents of LED BLUs

6.3.1 Light Guides

The latest trends in the basic constituent members of LED backlights, especially light guides, reflector sheets and LEDs are described below.

The basic properties required for the light guides include: (1) high transparency with no inclusion of foreign bodies, (2) no yellowing of the resin from which the light guide is made, (3) a precise shaping (no pulling or warping), (4) no residual stress and (5) superior mechanical properties. In addition to these requirements, there is a strong demand for the light guides to be thin and lightweight. The thinnest light guide used in the LED backlights currently on the market is 0.4 mm thick. Making thinner light guides involves the serious consideration of various issues. These include: reduced coupling efficiency since the incident light aperture is made smaller than the LED module, increased cost since greater care is needed for manufacturing, and increased difficulty in handling during assembly and inspection. Acrylic resin materials are popular choices in terms of their durability and cost. To achieve a reduction in weight, trials have been conducted to use amorphous cycloolefin-based resins having low specific gravity. The properties of these resin materials are compared in Table 6.1. Amorphous cycloolefin-based materials such as Zeonor 1060R and 1430R have properties such as high transparency and low absorption that are suitable for light guides, but suffer from problems such as requiring careful management due to their tendency to become denatured by oxidization in the molding process when in pellet form, and requiring careful handling due to the formation of cracks with oils.

6.3.2 Reflector Sheets

The reflector sheet is disposed at the lower surface of the light guide (Figure 6.1) and has the function of reflecting light that has escaped from the light guide back to the light guide again. Materials used for the reflector sheet include white PET which has diffusive properties, evaporated metal films such as aluminum which have specular reflection properties, and laminated films of polyester resins. The laminated films have the greatest reflectance,

Table 6.1 Comparison of the properties of resin materials.

	Units	Zeonor 1060R	Zeonor 1430R	PC	PMMA
Ray transmittance	%	92.3	92.0	90.9	92.8
Haze	%	0.34	0.30	0.76	0.40
Glass transition temperature	°C	100	136	140	100
Moisture absorption rate	%	0.002	0.002	0.15	0.27
DuPont impact strength	J	26.2	22.0	26.8	0.3
Rockwell strength		55.4	80.1	78.5	97.2

which is about 95% in the wavelength region from 400 nm to 800 nm. Although the laminated polyester reflector sheet is beneficial in terms of increasing the backlight brightness, it is relatively expensive and thus its use is limited to high-brightness backlights at the higher end of the market. A thickness of 60 μm has been achieved. The thickness of the prism sheet is about 110 μm, but there are plans to reduce this to 65 μm.

6.3.3 LEDs

The demands placed on the LEDs include high light emission efficiency, high brightness and reduced cost. The efficiency of white LEDs is about 50 lm/W, but plans have recently been unveiled to start shipping 100 lm/W LED samples which surpass the light-generating efficiency of the ordinary fluorescent lamps. It is expected that these devices will appear in commercial products within 4–5 years.

6.4 Various LED Backlight Configurations

LED backlights can be categorized into multiple light source configurations in which a plurality of LED modules are arranged at the end surface of the light guide, and point light source configurations in which there is only one light source. Backlights with multiple light sources can further be divided into dual prism configurations (like the one shown in Figure 6.1) where there are two prism sheets with prisms formed in orthogonal directions, and reverse prism configurations[2] like the one shown in Figure 6.6. The backlight has a single prism sheet with the prism surface facing downwards (towards the light guide). Table 6.2 lists characteristics of these configurations with regard to achieving high brightness, uniform intensity and thin backlight structure. There is little difference between them in achieving increased brightness. The dual prism configuration is particularly good at

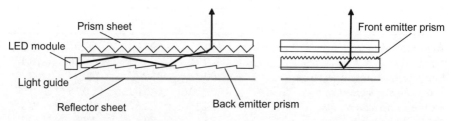

Figure 6.6 Structure of a reverse prism configuration.

Table 6.2 Comparison of LED backlight configurations.

Configuration	Multiple light source		Point light source
	Dual prism	Reverse prism	
Increased brightness	Satisfactory	Good	Satisfactory
Uniform intensity	Good (Industry standard)	Poor: Eyeball effect occurs near LEDs	Poor: Liable to contain bright lines and color fringes
Thin structure	Poor: Two prism sheets	Satisfactory: One prism sheet	Good: No prism sheet

achieving uniform brightness. Producing a thin backlight structure depends on reducing the thickness of each constituent member. The properties of the reverse prism configuration complement those of the dual prism configuration and are effective for achieving high brightness and a thin structure. It can be made thinner because there is only one prism sheet, and it can achieve an equivalent increase of brightness by collimating the light radiated from the prisms. Point light source configurations are particularly effective in achieving thinner structures. There are no configurations that simultaneously meet all the requirements for high brightness, uniform intensity and a thin structure.

6.4.1 Reverse Prism Configuration

Compared with the dual prism configuration of Figure 6.1, the reverse prism configuration can achieve better brightness in the forward direction by controlling the intensity of light emitted from the prisms, and can be made thinner because it uses only one prism sheet. Figure 6.6 shows a structure of the reverse prism configuration, explaining how the emission intensity is controlled. Back emitter prisms are formed on the back surface of the light guide, and form low-rise structures in which the angle of the back prisms is only a few degrees. Front emitter prisms are formed at the right angle to the back prisms on the front side of the light guide. The angle of the front emitter prisms is set to about 40°. As mentioned above, the prism surface of the prism sheet is facing downward, thereby forming an asymmetrical structure, unlike the dual prism structure.

In the reverse prism configuration, the emission angle distribution is controlled by the low angles of the back emitter prisms and front emitter prisms.

Light entering from the end surface of the light guide hits the back emitter prisms. Since the back emitter prisms have a low angle, light is emitted from the light-guide surface at a relatively shallow angle. The angle of this emitted light is transformed to the z-direction by the reverse prism, and the light is emitted from the prism sheet. Since the distribution of angles at which light is emitted from the light guide is more or less preserved, the result is that the light is condensed in the y–z-direction. The emission angle distribution in the z–x plane is controlled by the condensing action of the front emitter prisms. As a result of these condensing effects in two directions, it can be seen that a stronger condensing effect is achieved ($14°$ in the y–z plane and $29°$ in the z–x plane) than with the dual prism configuration ($26°$). This makes it possible to achieve brightness in the forward direction 2.4 times greater than that of the dual prism configuration.

One of the issues affecting this configuration is uniformity of the emission intensity. In particular, it is important to eliminate the non-uniformity of intensity in the vicinity of the LED modules (the so-called 'eyeball effect'). In the z–x plane close to the LED modules, most of the light is reflected by the front emitter prisms and returns to the light guide since the light is incident on the front emitter prisms at a shallow angle relative to the z–direction. Consequently, the emitted brightness is greater in the vicinity of the LED modules than between the LED modules. To correct this non-uniformity, the back emitter prism pattern is modified to a more diffused pattern near the LED modules.

6.4.2 Point Light Source Configuration

With the point light source configuration, an increase in brightness in the forward direction can be achieved by controlling the emitted light intensity in the same way as in the reverse prism configuration. A thinner structure can be achieved by implementing condensing functions on two axes, allowing the elimination of the prism sheets.[3,4] Figure 6.7 shows a point light source configuration, explaining how it condenses the emitted light. The LED backlight unit consists of three parts: an LED module, a light guide and a reflector sheet. The prism sheets and diffuser sheets are not necessary. The LED module is situated at a corner of the light guide and is equipped with a two-chip LED device inside the package. A reflector sheet is provided at the back surface of the light guide and provides a function of reflecting light that has escaped from the back surface of the light guide. The back emitter prisms are arranged in concentric circles centered at the position of the LED module to allow the emission intensity distribution to be controlled by the radial

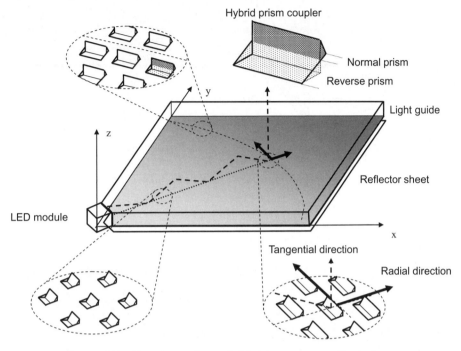

Figure 6.7 Structure of a point light source configuration.

distance. In this way, light entering the light guide from the LED module in any direction falls onto the back emitter prisms at a right angle in the x–y plane. Thus the emission angle in the radial direction is correspondingly more or less perpendicular. A uniform emission intensity distribution in the radial direction can be realized by curving the back emitter prisms.

Figure 6.8(a) shows a cross section of the back emitter prism. Light falling on the back emitter prism includes Mode 1 which reaches the back emitter prism by total internal reflection. Mode 2 is reflected off the bottom surface of the light guide within a distance d_1. Modes 1 and 2 are converted from the conduction mode to the emission mode by the back emitter prism and are emitted out of the light guide. Since there is a large angle component that is fully reflected by the back emitter prism, the emission angle distribution is broader.

To produce a condensing action in the radial direction, the double prism of Figure 6.8(b) is used as the back emitter prisms. The double prism is provided with a reversed prism structure in the region up to the distance d_1. The angles at which Mode 2 light is incident on the convex prism is confined

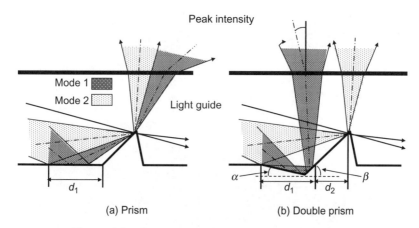

Figure 6.8 Cross section of a back emitter prism.

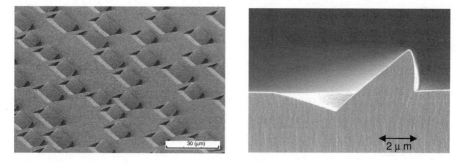

Figure 6.9 Double prism arrangement and cross-sectional view.

to angles corresponding to the low angle of the convex prism, and these light rays are converted into the region of Mode 1. In this way, the light is condensed in the radial direction. Combined with the condensing effect of the concentric arrangement, condensing on two axes is provided without the use of the prism sheets.

Uniformity is an important issue for the point light source configurations. Color fringing occurs due to the refraction originating from the concentric arrangement of the back emitter prisms, and moiré patterns are caused by interference with the pixels of the liquid crystal display panel. These issues can be overcome by randomly arranging the back emitter prisms in the radial direction within a certain dispersion width. Figure 6.9 shows the double prism arrangement and a cross section of the double prism.

6.5 Conclusions

The latest trends in LED backlight units with regard to dual prism sheets, reverse prisms and point light source configurations have been discussed. The essential function of LED backlights is to spread out the light from a point-source to a two-dimensional surface. With future advancements in micromachining and design technology, it should become possible to develop LED backlights that satisfy increased brightness, better uniformity and thinner structure.

References

[1] Fuji Chimera Research Institute, Inc. (2006) Report on the future prospects of next-generation mobile phones and key devices.
[2] Kalantar, K. *et al.* (2004) 'Backlight Unit with Double-surface Light Emission using a Single Micro-structured Lightguide Plate', *J. Soc. Information Display*, Vol. 12, pp. 379–388.
[3] Funamoto, A. *et al.* (2004) 'Prism-sheetless High Bright Backlight System for Mobile Phone', *Proc. of IDW '04*, pp. 687–690.
[4] Funamoto, A. *et al.* (2005) 'Diffusive-sheetless Backlight System for Mobile Phone', *Proc. of IDW' 05*, pp. 1277–1280.

Part Two
Light Source Devices

7

CCFL Backlights

K. Yamaguchi

Panasonic Photo & Lighting

7.1 Introduction

Cold cathode fluorescent lamps (CCFLs) have been widely used as light sources for LCD backlight units and they will continue playing an important role as light sources in the future. The major reasons for their wide acceptance are their ease of use, long life and various other advantages over other light sources. In thischapter, the fundamental technologies and likely future trends in CCFLs will be discussed.

7.2 Structure and Operating Principle of CCFLs

7.2.1 Basic Structure

Figure 7.1 shows the typical structure of a CCFL. The inner wall of a cylindrical glass tube is covered with phosphor, a pair of electrodes are sealed at both ends of the tube, and Ar and Ne are admitted with a small amount of Hg. Electrodes are a cup-shaped hollow type. The shape suppresses the effective sputtering rate caused by energetic ion bombardments. The electrode material is typically Ni, but high-melting point metals such as Mo and Nb are also used to allow larger discharge currents and longer life. Electrical conduction is made by hermetically sealed electrode leads.

LCD Backlights   Edited by Shunsuke Kobayashi, Shigeo Mikoshiba and Sungkyoo Lim
© 2009 John Wiley & Sons, Ltd.

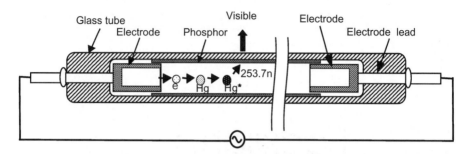

Figure 7.1 The structure and light emission of a CCFL.

The content gas is Ne with 3–10% Ar at a total pressure of 11–12 kPa for note PCs, and 7–9 kPa for monitors/TVs. The amount of Hg is several mg or less. Rare earth phosphors of YOX (Y_2O_3:Eu^{3+}) for red, LAP (LaPO$_4$:Tb^{3+}) for green and BAM ($Ba_2Al_{16}O_{27}$:Eu^{2+}) for blue are commonly used for Hg lines. The tube is made of hard glass which has a high transparency. Electrode leads are kovar or tungsten which have a thermal expansion coefficient close to that of hard glass. The tube length varies from 50 mm to 1600 mm, the outer diameter from 1.6 mm to 4.0 mm, and the inner diameter from 1.2 mm to 3.0 mm.

7.2.2 Operating Principle

When a starting voltage is applied across the electrodes, free electrons in the tube are accelerated and ionize neutral gas atoms. Positive ions thus created travel to the cathode and form a cathode fall region where an intense electric field exists. The ions bombard the cathode to emit secondary electrons. The electrons are accelerated towards the anode and collide with neutral atoms to ionize them. After the transition period, a self-sustained discharge with an operating voltage (lamp voltage) lower than the starting voltage is established. With the application of an AC voltage, electrodes act as cathodes as well as anodes depending on the polarity. Nevertheless, since the operating frequency is high at several tens of kHz, application of the starting voltage is not required except for the initial turning on. The region of the discharge other than the cathode fall is called the positive column. In the positive column, Hg atoms are excited by electronic collisions and emit vacuum ultraviolet (VUV) radiation of 253.7 nm. The VUV hits the phosphor to emit visible radiation.

The drive circuit of the lamp is called an inverter in which an internal or an external resonance circuit is used. The output voltage waveforms are close to sinusoidal. One of the requirements for the inverter is that the open-ended transformer output voltage should be higher than the starting voltage of the lamp. The drive frequency is 40–150 kHz for short discharge lamps which are used, for example, in digital cameras and 40–65 kHz for long discharge lamps which are used, for example, in monitors/TVs. For short lamps, a high voltage is applied to one of the electrodes while the other electrode is grounded. For some of the long lamps, high voltages of opposing polarities are applied to the two electrodes. Generally an inverter is required for each lamp. An inverter circuit in which one transformer can drive two lamps is becoming common for TVs and monitors.

7.3 Basic Characteristics of CCFLs

7.3.1 Electrical Properties

7.3.1.1 Dark-room Ignition

As explained in the previous section, free electrons are required to initiate the discharge. Also the number of the free electrons affects the delay of ignition. If an LC-TV is left 'off' for a prolonged length of time and kept in a dark room, then the delay of the ignition is extended. In order to shorten the delay to within 1 s for TVs or 0.5 s for monitors, an electron emitter is provided in the vicinity of the electrodes.

7.3.1.2 Starting Voltage Dependence on Lamp Length and Diameter

The starting voltage, V_s, is the value required to ignite a lamp which has been kept in a bright room, and depends on the lamp structure as well as the ambient temperature. As shown in Figure 7.2, V_s becomes higher as the lamp becomes longer and thinner due to the reduction of the electric field and the increase of diffusion loss of charged particles to the tube wall. V_s is also reduced as the ambient temperature becomes higher (Figure 7.3). This is due to an increase in the mercury vapor pressure. Also V_s becomes higher as the Ar content is increased.

7.3.1.3 Lamp Current versus Lamp Voltage

Figure 7.4 indicates that the lamp voltage is reduced as the lamp current is increased, showing a negative resistance characteristic. This is because the axial electric field in the positive column becomes weaker with an increase

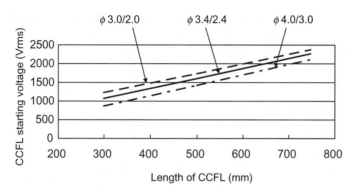

Figure 7.2 Starting voltage versus lamp length and diameter.
Driving: high voltage applied to one of the electrodes;
Lamp length: 300 mm;
Gas: Ne + Ar(5%), 8 kPa (60 Torr);
Electrode: Ni;
Ambient temperature: 0°C.

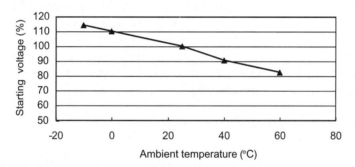

Figure 7.3 Starting voltage versus ambient temperature.
Driving: high voltage applied to one of the electrodes;
Lamp length: 300 mm;
Lamp diameter: 3.0/2.0 mm;
Gas: Ne + Ar(5%), 8 kPa (60 Torr);
Electrode: Ni.

in the current. Hence, even with a weaker electric field, the number of ion-izing collisions is sufficient to sustain the discharge at high lamp current levels. With smaller diameter tubes, ion–electron recombination at the glass tube wall increases. In order to compensate for this loss, a higher elec-tric field is required to sustain the discharge, resulting in a higher lamp voltage.

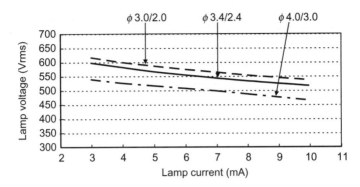

Figure 7.4 Lamp current versus lamp voltage.
Driving: high voltage applied to one of the electrodes;
Lamp length: 300 mm;
Gas: Ne + Ar(5%), 8 kPa (60 Torr);
Electrode: Ni;
Ambient temperature: 25°C.

7.3.2 Optical Properties

7.3.2.1 Luminance

Luminance increases as the lamp current is increased and the tube diameter is reduced (Figure 7.5). The peak luminance is obtained at an ambient temperature of 30–40 °C as can be seen in Figure 7.6. The mercury vapor pressure is determined by the coldest spot of the lamp. A decrease in the luminance at elevated ambient temperatures is caused by imprisonment of the 253.7 nm Hg resonant line which becomes dominant when the Hg vapor pressure is too high. Also the electron energy is reduced. Thus, driving the CCFLs at an optimum lamp temperature is important.

7.3.2.2 Chromaticity

Reproducible colors with the RGB phosphors are expressed by the region inside the triangle of the CIE xy chromaticity diagram of Figure 7.7. The area for conventional CCFLs (labeled 'Old') is 70% of the area defined by the National Television Standards Committee (NTSC). The recent phosphor (labeled 'New') provides an area wider than 90% of the NTSC.

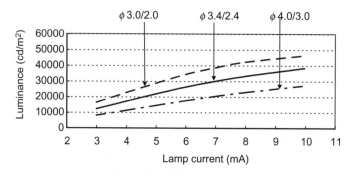

Figure 7.5 Lamp current versus luminance.
Driving: high voltage applied to one of the electrodes;
Lamp length: 300 mm;
Gas: Ne + Ar(5%), 8 kPa (60 Torr)
Electrode: Ni;
Ambient temperature: 25 °C;
Color temperature: 9100 °C;
Measured point of luminance: center of lamp.

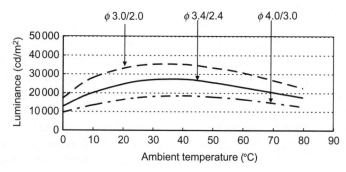

Figure 7.6 Ambient temperature versus luminance
Driving: high voltage applied to one of the electrodes;
Lamp length: 300 mm;
Gas: Ne + Ar(5%), 8 kPa (60 Torr);
Electrode: Ni;
Color temperature: 9100 °C; Measured point of luminance: center of lamp.

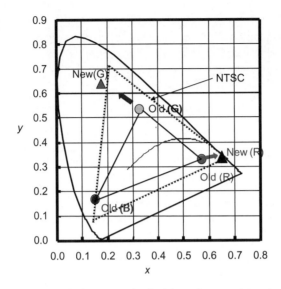

Figure 7.7 Color gamut of old and new phosphor.

7.3.3 Lamp Life

7.3.3.1 Luminance Life

The lamp life is defined as the time when luminance is reduced to 50% of the initial value. Figure 7.8 shows luminance degradation of a lamp. There are two major factors which dominate the luminance life. One is the degradation of phosphor which is caused by irradiation of 185 nm VUV radiation from the excited Hg. The other is a loss of transparency due to contamination of the phosphor surface by Hg. The transparency of the glass tube is degraded due to solarization by VUV irradiation and adsorption of Hg. Both these factors can be reduced to some extent by coating a protecting layer on the surface of the phosphor particles and on the inner wall of the tube. The life is also shortened by the loss of gases. During the sputtering of the cathode, mercury atoms are buried into the glass tube together with the sputtered materials.

7.3.3.2 Electrode Life

Sputtering of the electrodes reduces the current emission from the electrodes and hence reduces the luminance. In practice, the electrodes are designed to have a longer life than the half-luminance life. In order to reduce the

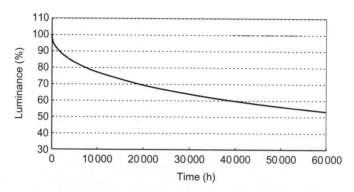

Figure 7.8 Degradation of lamp luminance.
Driving: high voltage applied to one of the electrodes;
Lamp length: 678 mm;
Lamp diameter: 3.0/2.0 mm;
Gas: Ne + Ar(5%), 8 kPa (60 Torr);
Electrode: Ni;
Ambient temperature: 25 °C;
Lamp current: 6 mA;
Measured point of luminance: center of lamp.

sputtering, it is effective to increase the total gas pressure as well as the Ar content. Also the use of metals such as Nb or Mo which have high melting points is preferable. As mentioned earlier, the lamp is usually driven with an AC voltage. If the voltage loses balance of the positive and negative polarities, that is if the positive period, for instance, is longer than the negative period, then this causes cataphoresis when ionized Hg atoms move towards one of the electrodes. Since Hg atoms are effective in reducing the sputtering rate, a smaller amount of Hg results in a shorter electrode life. Also, if there is a temperature gradient in the lamp, Hg atoms are not uniformly distributed in the tube, which again causes heavier sputtering.

7.3.3.3 Other Factors Effecting the Lamp Life

Another factor which influences the lamp life is a change in color, caused by differential aging of the RGB phosphors. Blue emitting BAM phosphor has a fast degradation and coating with La may be effective in extending the life. Other than the lamp life, yellowing of the resin used in the backlight unit occurs due to 313 nm UV radiation from the lamp. The use of a 313 nm-cut filter, or the use of glass that does not transmit that line, is effective.

7.4 Future Trends in CCFLs

7.4.1 Longer Lamps

Larger diagonal displays will become a main stream of LC-TVs, requiring longer CCFLs. For a 65-inch diagonal display with an aspect ratio of 16:9, the lamp will be 1450 mm long. In order to avoid a non-uniform temperature distribution, it is desirable to place the lamps horizontally. Due to a limitation of manufacturing facilities, however, the lamps may have to be placed vertically instead of making longer lamps. In such a case, special measures have to be taken to make the lamp temperature uniform.

Long lamps have the following technical problems:

(1) uneven thickness of the phosphor layer along the axial direction;
(2) uneven density of phosphor along the axial direction
(3) uneven distribution of Hg vapor along the axial direction.

These result in a build-up of luminance, color and discharge non-uniformities. Items (1) and (2) are caused by the specific gravity and the shape of the phosphor particles, for which an improvement in the coating technique is required. Another problem for the long lamps is that a higher starting voltage is required.

7.4.2 Wider Color Gamut

Improvements in RGB phosphors are essential for achieving a wider color gamut. Shifting the red peak to a longer wavelength, and green and blue peaks to shorter wavelengths favors the widening of the color gamut. However, there is a loss of luminance associated with this since visual sensitivity becomes lower in the infrared and ultraviolet regions. Accordingly improvements to the phosphor quantum efficiencies of saturated colors are also necessary. The lamp emission spectra and color filter transmission spectra should be matched. For this purpose, RGB lamp emission with a single peak for each color is desirable.

7.4.3 Higher Luminance

LCD luminance of $250 \, \text{cd}/\text{m}^2$ in 2001 was increased to $700 \, \text{cd}/\text{m}^2$ in 2006; nevertheless a 20% higher luminance is still demanded. To increase the luminance, increasing the lamp current as well as reducing the gas pressure is effective, although lamp life is reduced due to increased sputtering. The development of cathode materials which have a lower sputtering rate with a high discharge current and a low gas pressure, is required.

7.4.4 Higher Luminous Efficacy

One of the methods for obtaining higher luminous efficacy is to increase the luminance under a constant input power, and the other is to reduce the input power while retaining the luminance. The efficacy improvement is especially important in realizing low-power devices. For LC-TVs, 70% of the power is consumed by the CCFLs.

7.4.4.1 Optimization of Gas Content

The nature of the positive column discharge is strongly affected by the gas. In the positive column, electronic collisions ionize neutral atoms. Both ions and electrons travel to the tube wall due to ambipolar diffusion and they are neutralized there. Also volume recombinations of the ions and electrons take place in the discharge space.. These energy transfer processes depend on the electron temperature and affect the luminous efficacy. The optimization of the gas composition and pressure should also take into consideration the lamp voltage and electrode sputtering. Present CCFLs adopt a gas pressure which is higher than the optimum value, indicating that there could be further optimization of the gas pressure.

7.4.4.2 Reduction of Electrode Loss

Electrode loss is defined as energy loss in the cathode fall region. The use of high emission type electrodes lowers the cathode fall voltage by 50 V, but with a sacrifice of life. Emitters using diamonds are being investigated.

7.4.4.3 Other Factors for Efficacy Improvement

The lamp voltage should be reduced further. The lamp voltage of a 47-inch LC-TV, for instance, is 1770 V when the lamp current is 6 mA. In order to reduce the lamp power by 10%, while keeping the current constant, a voltage reduction of 170 V is required. This may be attained by reduction of the electrode loss and optimization of the gas content as mentioned above.

The use of a U-shaped lamp is another way in which the number of CCFLs, and hence the electrode loss, is reduced by a half. However, fabrication, as well as the ignition, of U-shaped CCFLs becomes difficult. A multiple number of external electrode fluorescent lamps (EEFLs) can be driven with one inverter. The lamp is operated at a low current, enabling higher efficacy to be achieved. A drawback of the EEFLs is their low luminous flux, so that a larger number of lamps are required.

7.5 Conclusions

CCFLs have long been used for LC backlights. There have also been – and there will continue to be – further improvements to the lamp characteristics. CCFLs will be used for low-power note PCs, high-end LC-TVs and also for various LC displays in the future.

8

CCFL Inverters

T. Uematsu

TDK

8.1 Introduction

Recently home-use electronic equipment using digital signal processing has become widely available, among them flat panel displays (FPDs). This chapter discusses technical issues and future trends in CCFL inverters for LC-TVs.

8.2 Various Drive Schemes of CCFL Inverters

Various drive schemes of CCFLs are depicted in Figure 8.1. For 40-inch diagonal LC-TVs, approximately 20 CCFLs are provided beneath the LC panel. For the pseudo U-shaped drive of Figure 8.1(a), two lamps are connected in series and driven together, similar to driving U-shaped lamps. For the one-side drive of Figure 8.1(b), the drive voltages are fed through one of the electrodes, while the other electrodes are grounded. The two-side drive of Figure 8.1(c), on the other hand, applies voltages of opposing polarities to the two electrodes. For inverters, the U-shaped drive, pseudo U-shaped drive and two-side drive can be treated identically.

Appropriate drive schemes are chosen by considering the diagonal sizes of LC displays. The U-shaped and pseudo-U-shaped drives are used for the 30-inch class, the two-side drive is used for the 40-inch class and above, and

LCD Backlights   Edited by Shunsuke Kobayashi, Shigeo Mikoshiba and Sungkyoo Lim
© 2009 John Wiley & Sons, Ltd.

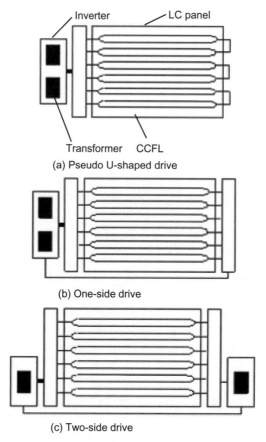

Figure 8.1 Various drive schemes.

the one-side drive is used for sizes between 30 and 40 inches. A choice is also made to reduce the inverter output voltage and to attain luminance uniformity along the tube axis. Figure 8.2 shows an example of voltage–current characteristics.[1] The diameter of the CCFL is 4 mm, the electrode material is Mo and the gas pressure is 8 kPa. The lamp lengths of 1100 mm and 300 mm correspond to diagonal sizes 50-inch and 14-inch, respectively. The lamp voltage increases approximately in proportion with the lamp length, requiring a higher insulating voltage and larger transformers.

The temperature distribution along the lamp axis is shown in Figure 8.3 for the lamps having 4 mm diameter and 8 kPa gas pressure when the one-side drive scheme is used.[1] The lamp lengths 800 mm and 600 mm

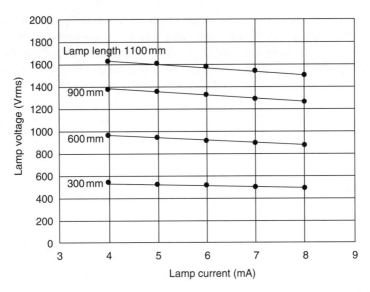

Figure 8.2 Voltage–current characteristics (reproduced by permission of Harison Toshiba Lighting Corp.).

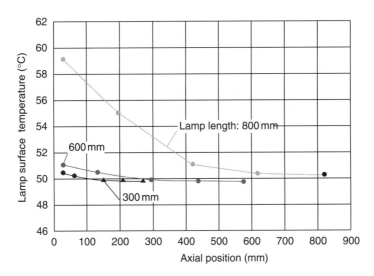

Figure 8.3 Axial temperature distribution of lamp surface; one-side drive with high voltage electrodes at the axial position 0 (reproduced by permission of Harison Toshiba Lighting Corp.).

correspond to 38-inch and 27-inch, respectively. As the lamp becomes longer, the temperature non-uniformity increases. Likewise, the luminance distribution along the axis increases with the lamp length due to the difference in Hg vapor pressure. By using the pseudo U-shaped drive or two-side drive, the drive voltage can be reduced to a half, while the luminance uniformity is improved. From Figure 8.3 it can be concluded that a 1200 mm long lamp (57-inch screen) will have relatively uniform luminance when the two-side drive scheme is adopted. The center of the tube will have the lowest luminance.

8.3 Equivalent Circuit of CCFLs

A voltage–current characteristic of a lamp used in 40-inch LC-TVs was measured and is shown in Figure 8.4. With a lamp current higher than 4.5 mA, the lamp exhibits properties of negative resistance. This implies that, if the backlight luminance is increased by controlling the lamp current, the lamp impedance shifts from positive to negative. Two current levels can be established under an identical voltage, indicating that the discharge is unstable. Discharge stability can be assured by connecting a ballast component in

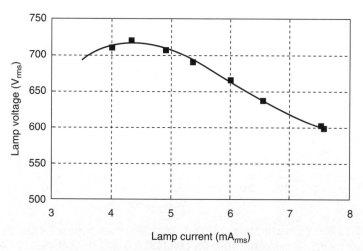

Figure 8.4 Voltage–current characteristic.

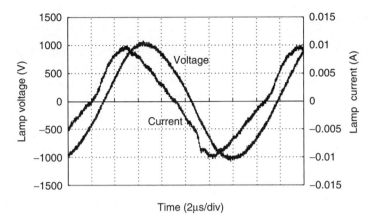

Figure 8.5 Voltage and current waveforms.

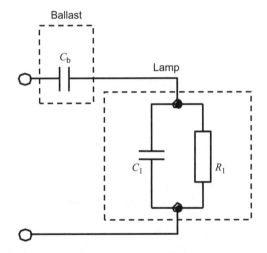

Figure 8.6 Equivalent circuit of a fluorescent lamp.

series with the lamp. A capacitor or an inductor is adequate as a ballast component.

Figure 8.5 shows the waveforms of the lamp voltage and current. Since the phase of the current is leading that of the voltage, the lamp is a capacitive load. An equivalent circuit of the fluorescent lamp can then be expressed as shown in Figure 8.6.[2,3] In the figure, C_b is the ballast capacitor. The values of R_l and C_l depend on the lamp voltage.

8.4 Inverter Circuits

There are (a) full-bridge circuits, (b) half-bridge circuits and (c) push-pull circuits for the inverters, as explained in Figure 8.7. The lamp impedance, Z_{lamp} in the figure can be expressed by R_l and C_l of Figure 8.6. The components C_6, C_b, and L_r of Figure 8.7, and C_l of Figure 8.6 make up a resonance circuit. L_r is usually the leakage inductance of the transformer. The input voltage of the inverters, V_{in}, is typically 24 V. The ratio of turns of wires, n, is typically 40 for the full-bridge circuit and 80 for the half-bridge circuit.

The operation of the full-bridge circuit will be explained in detail. There are four switching elements (MOS FETs) S_1, S_2, S_3 and S_4 which operate according to Figure 8.8. The voltage difference across the points A and B, v_{AB}, is also indicated. The frequency is typically 40–60 kHz. An equivalent circuit to the inverter together with the lamp is shown in Figure 8.9. The designing of the inverter is done by considering the transfer function of the lamp current i_l and lamp voltage v_{AB}. A Bode plot of the circuit is shown in Figure 8.10. In this figure the gain represents a ratio of lamp current i_l to the input voltage v_{AB}. When the inverter input v_{AB} is 24 V as described above, then i_l is several mA, yielding gain of approximately –60 dB. The phase of the lamp current always leads the lamp voltage. As found from the gain plot, the lamp is a resonator. Although the circuit is driven with the square waves of Figure 8.8, the resultant output voltage and current at or near the resonance frequency become sinusoidal as shown in Figure 8.5.

Since the inverter is a resonant circuit, the lamp current and lamp luminance can be adjusted by controlling the frequency. When the inverter is driven at a frequency lower than the resonance of Figure 8.10, then the lamp current can be reduced by lowering the frequency. This is called pulse frequency modulation (PFM). Nevertheless PFM is not used for LC-TVs since sometimes the noise arising from the transformer vibration becomes audible. For LC-TVs, pulse width modulation (PWM) is utilized. The PWM controls the 'on' periods of the switching elements of Figure 8.8, by varying the duty ratio, d, while keeping the frequency constant. Here the duty ratio is defined by the time the voltage is applied during the allotted period as explained in Figure 8.11(a). Figure 8.11(b) shows the relation between the duty ratio and the amplitude of the fundamental frequency component which is obtained after expanding the voltage waveform into the Fourier series. If the lamp is operated at or near the resonance frequency, the higher-order components are filtered and the fundamental component remains. Figure 8.11(b) indicates that the lamp current can be controlled by adjusting the duty ratio.

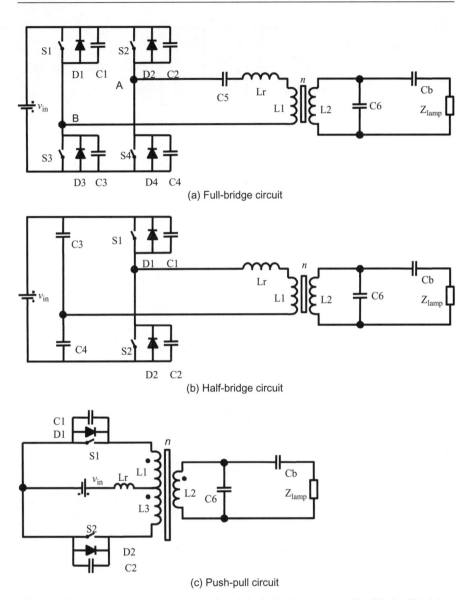

(a) Full-bridge circuit

(b) Half-bridge circuit

(c) Push-pull circuit

Figure 8.7 Various inverter circuits: (a) full-bridge circuit, (b) half-bridge circuit, and (c) push-pull circuit.

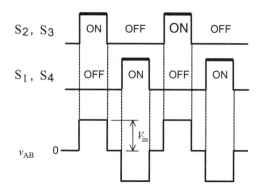

Figure 8.8 Switching sequence and v_{AB}.

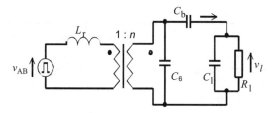

Figure 8.9 Equivalent circuit of an inverter and a lamp.

8.5 Driving of CCFLs with Inverters

Figure 8.12 shows various drive methods of large-area panels for the one-side drive. Figure 8.12(a) is a one-lamp drive in which each inverter drives a lamp. When the number of lamps is increased, the required number of inverters also increases. An advantage of this scheme is that the luminance of each lamp can be adjusted to obtain uniform luminance. Figure 8.12(b) is a multiple-lamp drive with a common switching circuit, in which one transformer is assigned to each lamp but there is only one switching circuit. The numbers of the switching elements and associated ICs can then be reduced. The lamp current, however, depends on the transformer characteristics which may not be identical. Figure 8.12(c) is a parallel drive of multiple lamps in which a transformer can drive a number of lamps. The scheme enables the number of the transformers to be reduced. Also since there is only one transformer for a backlight unit, the lamp current variation due to the transformer is eliminated. The performances of lamps, on the other hand, should be made identical.

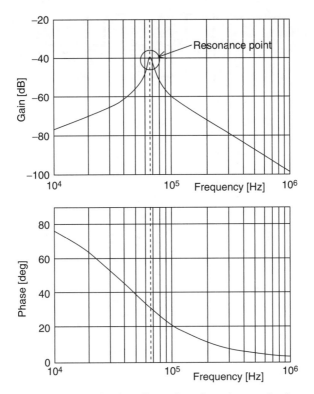

Figure 8.10 Transfer function of an inverter and a lamp.

8.6 Lamp Current Balancers for Driving Multiple Lamps

When driving a number of lamps with an inverter or with a transformer, the lamp current differs for each lamp, resulting in non-uniform luminance and also causing differential aging. This adds further requirements to the current-balancing circuit. Available options for the current balancers are

(a) the differential transformer scheme (ZAULaS),[4]
(b) the ring balancer scheme,[5] and
(c) the LC ballast scheme.[6]

Their circuit diagrams are shown in Figure 8.13. In the diagrams, v_{Tout} is equal to the voltage across C_6 of Figure 8.9. For all the schemes, the balancers also act as ballasts. Since the output voltage needs to be as high as 1000 V, only passive elements are used instead of active elements such as MOS FETs.

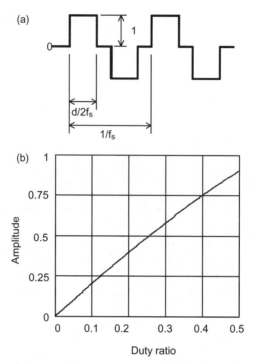

Figure 8.11 (a) Duty ratio and (b) amplitude of fundamental frequency component.

In the scheme of Figure 8.13(a), two lamps are driven with a differential transformer which makes the current equal for the two lamps. By connecting a differential transformer in a cascade fashion, the current of all the lamps can be made identical. The number of transformers required for the scheme is equal to the number of the lamps minus one and the scheme is alternatively called the tournament scheme. Unless the number of the lamps is equal to 2^n (where n is an integer), special means of driving have to be considered. Also n kinds of differential transformers have to be provided.

The ring balancer scheme is shown in Figure 8.13(b). A transformer is needed for each lamp, but the secondary circuits are connected in series in the form of a ring. Since the secondary current i_{loop} is identical to all the transformers, the primary lamp currents i_{lamp} also become identical. In order to reduce the lamp current, several hundreds of mH are required as a self inductance of the primary circuit. In order to reduce the size of the balancer transformers, the use of a high permittivity core as well as high coupling of

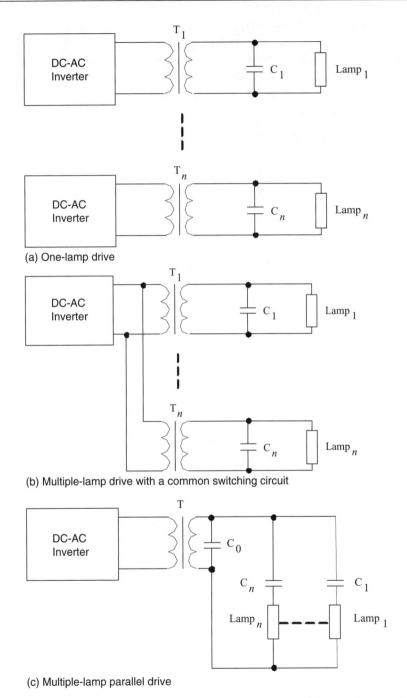

Figure 8.12 Drive methods for large-area panels: (a) one-lamp drive, (b) multiple-lamp drive with a common switching circuit, and (c) multiple-lamp parallel drive.

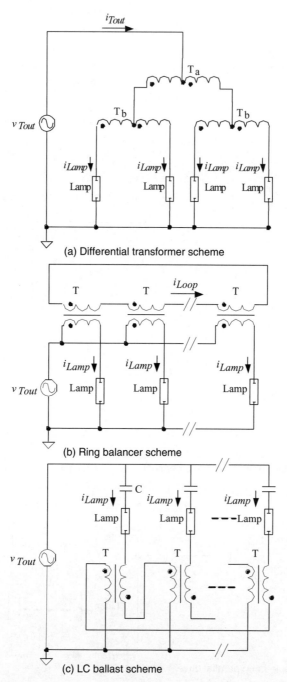

Figure 8.13 Various lamp current balancers: (a) differential transformer scheme, (b) ring balancer scheme, and (c) LC ballast scheme.

the primary and secondary windings are desirable. The ring balancer occupies the smallest area among the three.

Figure 8.13(c) shows an LC ballast scheme which consists of ballast capacitors and differential transformers. The variation of the lamp current can be suppressed as in the case of the differential transformer scheme of Figure 8.13(a). The difference compared with Figure 8.13(a) is that the voltage applied to the differential transformers can be reduced by the ballast capacitors.

8.7 Conclusions

There are many other techniques with regard to inverters. Digital control of inverters allows dynamic brightness control and dynamic color control. Picture quality improvement and reduction of energy consumption can further be achieved. The biggest issue for the inverters would be the cost.

References

[1] http://www.htl.co.jp/p_pro_04.html
[2] Kim, C.-G. *et al.* (2005) 'Modeling of CCFL using Lamp Delay and Stability Analysis of Backlight Inverter for Large Size LCD TV', *IEEE APEC '05*, pp. 1751–1757.
[3] Wu, T.-F. *et al.* (1997) 'A Spice Circuit Model for Low-Pressure Gaseous Discharge Lamps Operating at High Frequency', *IEEE Trans. on Industrial Electronics*, **44**, pp. 428–431.
[4] Ushijima, M. (2004) 'Inverter circuit for discharge lamps for multi-lamp lighting and surface light source system', US Patent #20040155596A1 (filed Feb. 9, 2004).
[5] Jin, X. P. (2004) 'Current sharing scheme for multiple CCF Lamp operation', US Patent #20050093471A1 (filed Oct. 5, 2004).
[6] Morita, S. (2006) 'Ignition circuit of discharge lamp', Japan Patent Open, Tokukai 2006-12659.

9

HCFL Backlights

A. A. S. Sluyterman

Philips Lighting

9.1 HCFL Light Source as a Member of the Fluorescent Lamp Family

The hot cathode fluorescent lamp (HCFL) is, like CCFL light sources, a low-pressure mercury discharge lamp. However, is has a different cathode resulting in different lamp properties. In a tubular lamp a gas discharge is created in which electrons are generated that excite mercury atoms, which in turn generate UV light, with dominant wavelengths of 185 and 254 nm. This UV light is converted into visible light by means of phosphor that is deposited on the inside of the glass envelope of the lamp. This is shown in Figure 9.1.

The phosphor is a mixture of at least three components, associated with the colors red, green and blue. By varying the composition of the phosphor – the ratio between the components – the emitted spectrum and its color point can be tuned. An example of the light spectrum of a lamp is given in Figure 9.2. When used in an LCD backlight, the ratio between the phosphor components determines, together with the color filters on the LCD panel, the color points of the primary colors of the display.

In addition to the saturated mercury system (the mercury is present under normal operating conditions in both liquid and vapor forms) there is an inert buffer gas. The inert gas, often argon or neon, has three functions:

LCD Backlights   Edited by Shunsuke Kobayashi, Shigeo Mikoshiba and Sungkyoo Lim
© 2009 John Wiley & Sons, Ltd.

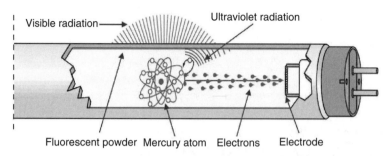

Figure 9.1 See-through view of an HCFL and the processes leading to visible light.

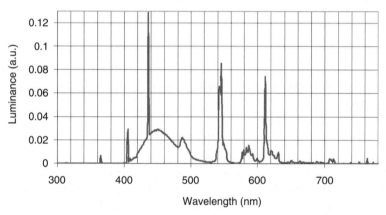

Figure 9.2 An example of the light spectrum of an HCFL light source used in an LCD backlight system.

(1) it reduces the speed of free electrons, as these electrons would cause ionization rather than excitation of the mercury atoms;
(2) it reduces the ion bombardment of the electrodes, and thus prolongs the life of the emitter material in the electrodes;
(3) it lowers the ignition voltage of the lamps.

The pressure of the noble gas is of the order of 500 Pa, depending on the lamp diameter among other things.

The pressure of the saturated mercury gas is about 1.2 Pa, depending on the lowest temperature inside the lamp. The pressure of the noble gases also influences the spatial distribution of the light that is emitted by the gas discharge itself. When a lamp is driven in DC mode, darker and brighter zones can be seen. The names associated with these zones are shown in Figure 9.3.

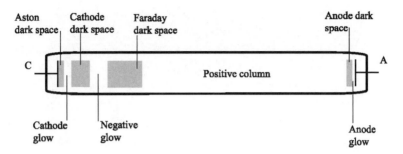

Figure 9.3 Illustration of the spatial distribution of the light generated by the gas discharge of the lamp, and the names associated with the darker and brighter zones.

Normally the lamps are operated in AC mode, and then the different zones cannot be seen anymore. In addition, the phosphor on the inside of the tube diffuses this spatial luminance distribution even further.

9.2 Introduction of the Hot Cathode in Fluorescent Lamps

For any gas discharge that is required to conduct a current it is essential that electrons can be extracted from electrodes. In cold cathode fluorescent lamps the process of secondary emission governs the emission of electrons. As a consequence of this process the lamp currents are typically limited to 10–20 mA. HCFL lamps have heated electrodes as shown in Figure 9.4. By heating the electrode, it is capable of emitting electrons through the so-called process of thermionic emission. The currents that can be emitted are typically between 50 mA and 1 A.

In an HCFL lamp, the electrodes are like a coil (Figure 9.4). The coils are made of tungsten and are coated with a suitable emitter material – a mixture of barium, calcium and strontium oxides. The coils are wound in such a way, that they can contain as much emitter material as possible and this material ensures that electrons are released when the coils are heated to about 1250 K. Prior to ignition of the lamps, the electrodes are preheated. Once the lamp is burning, the coils are kept at a steady temperature by:

(1) ohmic heating by lamp current,
(2) bombardment of the electrodes by fast ions emanating from the discharge,
(3) additional heating of the coils.

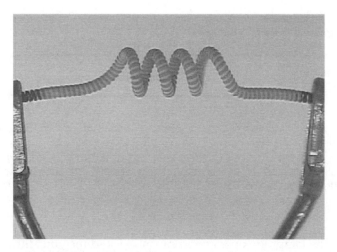

Figure 9.4 Electrode of an HCFL lamp.

The latter will be explained in the lamp driver section of this chapter.

Due to the differences in currents that can be emitted, CCFL lamps have small diameters (typically 4 mm) and HCFL lamps have larger diameters (typically 16 mm). Due to this diameter difference the lamp voltages are also significantly different. For CCFL lamps, the lamp voltages are of the order of 1500–2000 V/m, whereas for HCFL lamps, voltages are typically 150 V/m. These lower driving voltages are very advantageous, because this means that the capacitive leakage of lamp current towards the environment of the lamp is negligible. Furthermore, as a consequence of the larger diameter, the light output for HCFL lamps is significantly higher than for CCFL lamps. Because the HCFL can be driven at much lower voltages than the CCFL, the HCFL is particularly advantageous for large LCD displays (32-inch and larger) because then long lamps are needed.

9.3 Driving the HCFL

When an HCFL is ignited, the gas discharge runs in the 'arc mode'. This means that the larger the current through the discharge, the lower the voltage across the discharge. This negative I–V characteristic is common for all discharge lamps and can be described in terms of a 'negative resistance' characteristic of the lamp. As a result the lamp cannot be connected directly to an electricity source with a fixed voltage, because this would cause a 'current run-away' problem. The most common solution to this problem is to drive the lamp via an inductance, because driving it via a large resistor would lead to large power losses in the lamp driving circuit. For general lighting applications, the lamps can be driven at mains frequencies. For LCD backlighting applications – and nowadays more and more for general lighting applications as well – the lamps are driven with a frequency of around 85 kHz. The basic diagram for driving an HCFL is given in Figure 9.5.

To ignite the HCFL, the frequency of the driving voltage is brought to the resonance frequency, determined by the inductance and capacitor of the driving circuit. Once the lamp is ignited the current through the lamp is controlled by variation of the frequency of the driving voltage. Prior to the ignition of the lamp, the electrodes have to be preheated. This is done by setting the driving voltage frequency higher than the resonance frequency. The impedance of the capacitor then reduces, the discharge is not established and all the current goes through the two electrodes in a series connection. By choosing the proper frequency, the right amount of preheating can be achieved.

In LCD backlighting applications, dimming of the lamps is done via duty cycle dimming, meaning that during a part of the time the lamps are switched off. Again by choosing the proper frequency during the off period of the lamp, the electrodes will receive the correct amount of current needed to maintain the electrodes at the right temperature. Once this preheating and additional heating of the electrodes has been achieved, the life of the electrodes can be long (see Section 9.4) and the backlight can be dimmed by 10% of the nominal value.

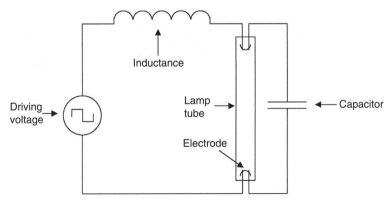

Figure 9.5 Basic diagram for driving an HCFL.

9.4 Cathode Life Properties of HCFL

During its life the emitter loses material due to evaporation and due to sputtering as a result of ion bombardment. When the electrodes are too hot, evaporation dominates, when the electrodes are too cold, sputtering dominates. Important factors in extending the life of the electrode are:

(1) the electrode design,
(2) the emitter volume,
(3) the preheating current – the current that is applied to the electrode in order to ensure a sufficiently high electrode temperature when the lamp is ignited, and
(4) the lamp current and (when present) the additional heating current.

For HCFL lamps the so-called R_{hot}/R_{cold} value is a measure of the electrode temperature. Here R_{cold} is the cold resistance of the electrode and R_{hot} is the resistance of the electrode when operated. For an optimal electrode operation, the R_{hot}/R_{cold} value has to be typically between 4.5 and 5. This is realized by the combination of the lamp current and the additional heating current. When the lamp is dimmed, the electrode temperature can be maintained by the additional heating current. In this way a good lifetime performance of HCFL lamps can be obtained, also for dimmed operation.

The lifetime of the cathode is determined by the consumption of the emitter material during its life. In particular, barium is of importance here as it guarantees good electron emission properties of the electrode. Tests have been carried out to determine the consumption of barium when the HCFL lamps are operated in the scanning mode.[1] After the HCFL lamps have operated for a specified period of time, the amount of barium on the electrode is determined by destructively analyzing the electrode. By means of an accelerated test, where the lamps were ignited three times per frame instead of once per frame, the barium consumption is determined for up to 12 000 simulated operating hours. The result of the test is given in Figure 9.6, which shows that virtually no barium is consumed during operation. So, based on the very limited consumption of barium, cathode life times far greater than 50 000 hours can be expected, once the optimal electrode operating conditions are fulfilled.

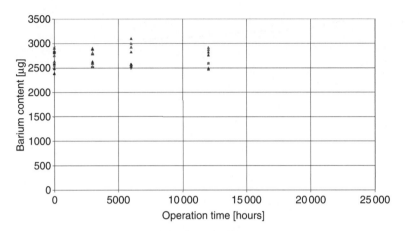

Figure 9.6 Barium content as a function of lamp operation hours; virtually no barium is lost at all.

9.5 Lumen Maintenance and Color Point Shift during Life

When using eight HCFLs with a diameter of 16 mm in a 32-inch backlight (see Section 9.6) the total phosphor surface is larger than in the case of 16 CCFLs with an internal diameter of 3 mm. As a result, the phosphor load is much lower and so less luminance deterioration and color shift can be expected. The luminance decay is given in Figure 9.7 and the color point shift is given in Figure 9.8. As a reference, typical data of CCFL lamps is given.[2]

These figures show that very good light output maintenance can be achieved with HCFL light sources – approximately 95% at 15 000 hours. Furthermore the color points shift by approximately $\Delta x = 0.008$ and $\Delta y = 0.01$ after 15 000 hours.

9.6 Designing a Backlight with HCFL

For a 32-inch backlight system, eight HCFLs are sufficient to give a light output of the LCD display of 550 cd/m² at a 35% duty cycle. Figure 9.9 shows a typical layout of such a backlight system. Incorporation of 16 mm diameter HCFLs into a light box with an internal thickness of 26 mm would, however, not give uniform illumination without additional measures since it would

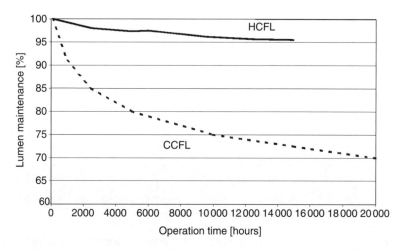

Figure 9.7 Luminance decay as a function of operating time.

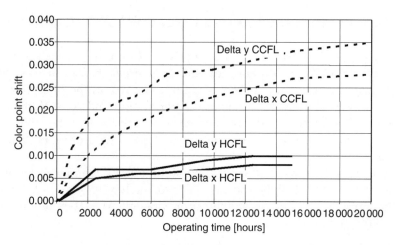

Figure 9.8 Color point shift as a function of operating time.

show bright stripes at the positions where the lamps are located. One option to solve this problem completely is a new optical construction using an optical coating on the light sources, as shown in Figure 9.10.[3] The construction consists of fluorescent lamps with a modified light distribution. In this case, the light flux emitted directly to the front face of the light box is redirected by a white, diffuse reflecting coating on top of the fluorescent tube.

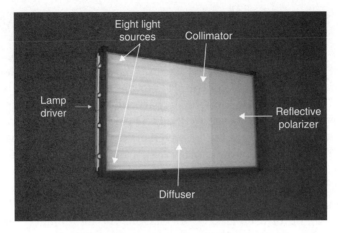

Figure 9.9 A see-through view of the 32-inch backlight system with eight fluorescent lamps, each with a diameter of 16mm.

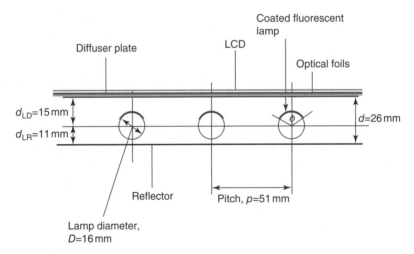

Figure 9.10 Backlight optical construction based on partially coated (over an angle ϕ) HCFLs.

The reflected light increases the light flux between two lamps. The light flux above the lamps and between the lamps can now be altered by changing the reflectivity of the coating and the angle over which the coating is deposited. This permits fine-tuning of the luminance uniformity and opens the way to fabrication of relatively thin and highly uniform backlights.

9.7 The Scanning Feature, Cost-effectively Enabled by HCFL

(1) In present day LCD TVs motion fidelity is still critical. Objects displayed on the screen become blurred as soon as they move over the screen. There are two potential contributors to this effect: The response time of the LCD panels can be too long, in particular for large screen sizes. This problem is being addressed by the LCD panel makers. They can use so-called overdrive to ease the inconvenience of the long response time.

(2) The hold-time effect of the display; when an image remains on the screen until it is refreshed, image smearing occurs in the human eye for moving objects that the viewer wants to follow. This second phenomenon can be tackled via the LCD backlight by reducing the illumination time of the backlight for each area on the screen. The illumination time can be expressed as a ratio, called the duty cycle of the frame period time.

Good motion fidelity leads to the concept of a backlight with a small duty cycle. However, a small duty cycle is not in itself enough; the backlight illumination phase (i.e. the illumination timing in relation to the writing sequence) is also important. A free-running backlight system with a small duty cycle will therefore not work. So from the timing point of view the LCD illumination must have a small duty cycle and must scroll down from the top of the screen to the bottom in phase with the addressing of the panel, called 'scanning backlight', as shown in Figure 9.11.

Figure 9.12 shows the optimum timing for a scanning backlight for large screen TV applications.[3] Basically, the illumination phase must be such that for each point on the screen the illumination is switched off, just before new video information is written into the panel. By using HCFLs there is so much luminance reserve that it is possible to use a duty cycle of 35% without sacrificing luminance. Since only eight HCFLs are needed an escalation of costs can be avoided.

9.8 The Dimming Feature

LCD panels have poor contrast in dark scenes because even in the dark state there is always some light leakage through the LCD panels. This is particularly noticeable when the panel is viewed at an angle. In dark scenes, the contrast of the displayed image can be improved by reducing the luminance of the backlight unit and at the same time increasing the panel transmission.[4] Dimming can be accomplished by a reduction of the illumination

Figure 9.11 Scanning backlight principle needed to obtain good motion representation in an LCD display, in which a light band scrolls over the screen. Each of the lamps is activated only for a short while.

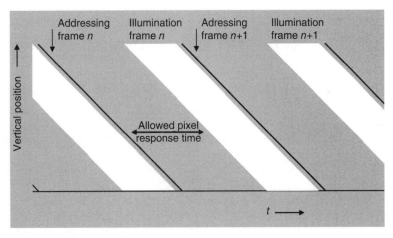

Figure 9.12 Ideal timing in a scanning backlight; the illumination and addressing of the panel vary in the same way with the vertical position on the screen.

duty cycle. In this respect, HCFL light sources are particularly well suited because the cathode heating can be controlled independently of the luminance of the lamp, so that dimming does not degrade the lifetime of the lamp. Using this system, luminance dimming by a factor of eight is possible.

In addition to the dimming benefits offered by HCFL light sources, the scanning backlight offers further benefits for dimming. With non-scanning dimmable backlights, there is no clear demarcation in illumination between frames so that a dimming action on one frame may spill over into neighboring frames. With a scanning backlight, however, because of the accurate in-phase timing of the illumination with the addressing of the panel, it is possible to address each complete frame independently of the previous or subsequent frames. This allows a much faster response to changes in illumination than is possible with non-scanning dimmable backlights, a feature that is particularly advantageous with transitions from dark to bright scenes.

9.9 Conclusions

HCFL light sources are very attractive for large size– easily up to 65-inch – LCD TV backlight applications because:

(1) they require a relatively low driving voltage,
(2) they give a surplus of light,
(3) they can be deeply dimmed to give a good dynamic contrast,
(4) they show a good lumen and color point maintenance, and
(5) they are very suitable for a scanning backlight for improved motion fidelity.

References

[1] Gielen, H. (2005) 'Life aspects of HCFL light sources in LCD backlight applications', SID '05 Digest of Technical Papers, pp. 1001–1003.
[2] Nishihara, T. and Takeda, Y. (2000) 'Improvement of lumen maintenance in cold cathode fluorescent lamp', IDW '00, pp. 379–382.
[3] Sluyterman, A. A. S. and Boonekamp, E. P. (2005) 'Architectural choices in a scanning backlight for large LCD TVs', SID '05 Digest of Technical Papers, pp. 996–999.
[4] Raman, N. and Hekstra, G. (2005) 'Dynamic contrast enhancement of liquid crystal displays with backlight modulation', IEEE Conf., Las Vegas, USA, pp. 197–198.

10

EEFL Backlights

J.-H. Ko

Hallym University

10.1 Introduction

An external electrode fluorescent lamp (EEFL) was originally designed and
used as a long-life light source at the early stage of research on plasma discharges. Now it is one of the major light sources for the backlight applications of LCDs. EEFL backlight technology was adopted for the first time in
2003 in 30-inch LC- TVs, and it has extended its available size in the range
23–42-inch TVs. EEFLs have been used mainly in the direct-lit backlight for
TV applications because multiple EEFLs can be operated in parallel by a
single inverter, an attractive feature for cost competitiveness of LCDs among
various flat panel displays. In addition, EEFLs have advantages such as
lower power consumption, longer lifetime, simpler backlight structure, etc.,
compared to the conventional CCFL backlight technology. The basic characteristics of the EEFL and recent technological trends and issues of EEFL
backlights will be summarized in this chapter. Future development targets
for EEFL backlights will also be briefly mentioned.

LCD Backlights   Edited by Shunsuke Kobayashi, Shigeo Mikoshiba and Sungkyoo Lim
© 2009 John Wiley & Sons, Ltd.

10.2 Basic Characteristics of EEFLs

10.2.1 Lamp Structure

The structure of an EEFL (Figure 10.1) is identical to that of a CCFL except for the electrodes. For the CCFL, hollow metal electrodes are inserted at both ends of the glass tube, while for the EEFL, external electrodes are located on the outer surface of the glass tube. The external electrode may be formed by inserting a metal cap on a pretreated glass surface, or by laminating on the glass surface a metal tape or foil beneath which a conductive resin is attached. Alternatively, dipping the end of the lamp tube into a metal paste, for example silver paste consisting of silver powder and a binder material, etc., may provide the electrode. The lamp tube is normally made of borosilicate glass containing low alkali content, and a tricolor phosphor is coated on the inner wall of the tube. The discharge space is filled with a mixture of inert gases such as neon and argon for Penning effects at a pressure in the range between 30 and 100 Torr, in addition to a minuscule amount of mercury as a source element of ultraviolet (UV) light. For the purpose of reducing mercury consumption and thus extending its lifetime, a protection layer such as yttrium oxide or aluminum oxide may be placed between the inner wall of the glass tube and the phosphor layer, or they may be dispersed on the surface of phosphor powders.

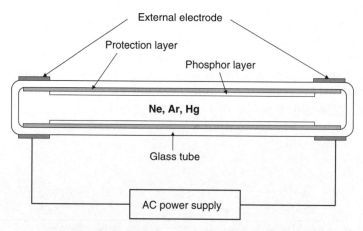

Figure 10.1 Structure of a conventional tubular EEFL.

10.2.2 Discharge Characteristics

The discharge induced in the EEFL is basically a dielectric barrier discharge (DBD). The glass tube plays the role of a capacitive dielectric barrier through which the displacement current passes by alternating the electric field. The impedance of the dielectric barrier limits the total current flowing through the discharge space, which indicates that additional ballast elements are not required for the operation of EEFLs. A glow discharge induced in EEFLs is sustained by the creation of charged particles due to ionizing collisions in the discharge and by secondary electron emission due to the ions colliding onto the inner glass wall underneath the electrodes. UV light is generated in the glow discharge by electronic impact excitations of mercury atoms. Generated UV light is transformed into visible light via the phosphor layer which plays the role of a wavelength converter. The spectral characteristics, corresponding color coordinates, correlated color temperature and color gamut of the output light are determined by the compositions and the mixing ratios of the tricolor phosphor materials.

Formation of wall charges is another archetypical characteristic of the EEFL. During the glow discharge, ions and electrons are accumulated on the inner surfaces of the glass tube underneath the electrodes. These wall charges terminate the discharge, because the polarity of the internal field induced by the wall charges is opposite to that of the externally applied field, reducing the effective voltage in the discharge space to a value below the minimum discharge voltage. These wall charges contribute to the formation of additional discharges when the externally applied voltage falls to zero and the effective voltage increases. This effect becomes more effective when an EEFL is driven with a square wave voltage having sharp pulse edges.

For EEFLs, the lamp current (I_L) increases as the lamp voltage (V_L) increases, showing positive resistance characteristics. This is due to the impedance of the glass tube wall. Figure 10.2 shows the lamp voltage as a function of the lamp current of an EEFL whose outer diameter, tube length and gas pressure are 2.6 mm, 420 mm and 60 Torr of Ne:Ar with a ratio of 9:1, respectively.[1] The I_L–V_L relationship for a CCFL with the same dimensions and gas conditions is also plotted for comparison. Contrary to the positive current–voltage characteristic of an EEFL, the CCFL shows a negative resistance characteristic, which needs a ballast element for the stabilization of the discharge. However, if the CCFL is connected to two ballast capacitors at both ends and if the voltage applied to the ballast capacitor is included into the lamp voltage, the I_L–V_L characteristic of the CCFL shows behavior similar to that of the EEFL including the threshold discharge current, the magnitude of the

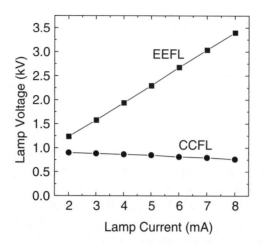

Figure 10.2 Relationships between lamp voltage and lamp current of a CCFL and a EEFL.[1] Both lamps have the same outer diameter, tube length and gas pressure which are: 2.6 mm, 420 mm and 60 Torr of Ne : Ar with a ratio of 9 : 1, respectively (reproduced by permission of the Society for Information Display).

firing voltage and the current increment with lamp voltage.[2] These results imply that the discharge mechanism of an EEFL is practically the same as that of a CCFL which has ballast capacitors at both ends.

The light-generating efficacy of the EEFL has also been shown to be comparable to that of a CCFL.[1] For EEFLs with an outer diameter of 2.6 mm and a tube length of 420 mm, the gas pressure for the highest efficacy was in the range between 40 Torr and 90 Torr under normal operating conditions. Several parameters affecting the luminance and efficacy of an EEFL will be discussed.

10.2.2.1 Lamp Current

Luminance increases as the lamp current increases in the low-current range but the rate of the increase becomes smaller with increasing current and then becomes saturated above particular current values.[1,2] This indicates that the efficacy becomes lower with increasing current density. This is due to elastic scattering losses and also excitations to useless nonultraviolet levels which increase with increasing current density. The rate of quenching collisions of the second kind by slow electrons, which take away the excitation energy from excited mercury, becomes larger than the rate of radiating transitions under high current density.[3]

10.2.2.2 Capacitance

Since the discharge in an EEFL is induced and maintained with the capacitive coupling through the glass wall, it is expected that the efficacy would depend on the capacitance of the glass wall located under the external electrodes. The capacitance also affects the amount of wall charges and the voltage distribution along the lamp tube. The applied voltage is divided into two parts, the voltage across the glass tube wall (V_G) at the electrode and the voltage of the glow discharge (V_P). As the capacitance increases, the ratio of V_P to the total applied voltage ($V_G + V_P$) increases,[4] which in turn increases the discharge energy dissipation and efficacy to some degree. This was demonstrated by investigating the effect of electrode length on the efficacy, which showed that the efficacy first increased and then saturated with increasing electrode length.[5,6] The same effects may be achieved by changing the thickness or the dielectric constant of the glass tube because the capacitance (C) is related to the dielectric constant (K), the thickness of the glass (d) and the electrode area (A) via $C = \varepsilon_0 KA/d$, where $\varepsilon_0 = 8.854 \times 10^{-12}$ F/m provided that the tubular glass is approximately equivalent to a parallel plate capacitor. In addition, the dielectric loss component of the glass material should also be considered for improving luminance and efficacy since the energy dissipation in the glass is proportional to the imaginary part of the complex dielectric constant of the glass capacitor.

10.2.2.3 Driving Voltage Conditions

There have been several studies reporting the dependence of the efficacy on the driving waveforms, in particular sinusoidal and square waves. Some studies reported that the efficacy becomes much higher when an EEFL is driven with a square wave with a duty ratio smaller than 100% compared to the case where it is driven with sinusoidal wave at the same input power and frequency.[7,8] Here, the duty ratio is defined by the ratio of the voltage-on time to the period of the driving voltage. If the duty ratio is smaller than 100%, double lightings are generated during a half period from the self-discharge due to the recombination of the wall charges which occurs when the externally applied voltage is eliminated. On the other hand, if the duty ratio of the square wave is 100%, the efficacy of the EEFL becomes lower than the case where it is driven by a sinusoidal wave in the frequency range below 150 kHz.[1] Combining these results, it is recommended that an EEFL should be driven with a square-wave voltage having a duty ratio smaller than 100%. Here sharp rising and falling edges of voltage pulses are required for the efficient double lightings.

It was shown that the efficacy increased with increasing driving frequency for both square-wave and sine-wave drives.[1] Since the impedance of the glass wall at the electrode is inversely proportional to the drive frequency, the voltage applied across the glass tube becomes low with increasing frequency and therefore the lamp voltage can be reduced. In addition, it is expected that the ionizing collision rate of the electrons will increase with increasing frequency, resulting in the efficient production of charged particles and thus a reduced cathode fall.[9] However, the leakage current to the metal chassis, as well as the possible generation of electromagnetic interference (EMI) generated by the high-frequency components of the square wave, should be considered when optimizing the waveform and the frequency.

10.2.3 Energy Balance of EEFLs

The luminous efficacy of EEFLs is reported to be comparable to that of CCFLs. Efficacy is obtained from the ratio of total luminous flux to the power consumed by the lamp. For a more complete understanding of the energy balance of an EEFL, it is necessary to measure the total UV output, which can be estimated indirectly from the procedure described by Mizojiri et al.[10] The luminous efficacy of the lamp η_L can be described by the equation:

$$\eta_L = \eta_{UV} \times \eta_E \times Q_L \times L_e, \tag{10.1}$$

where η_{UV}, η_E, Q_L and L_e are the UV generating efficiency, the energy conversion efficiency from UV to visible light, the average quantum efficiency of phosphors and the lumen equivalence of the emitting spectrum in units of lm/W, respectively.[10] If visible light is emitted from the discharge plasma, this component has to be added to the output flux. Luminous efficacy of the visible component in the typical low-pressure mercury fluorescent lamps is of the order of 1–2 lm/W.

The quantitative analysis of the energy balance of an EEFL has been reported elsewhere.[11] For the estimation of η_{UV}, the electro-optical characteristics of a standard EEFL for a 32-inch backlight unit were investigated. The outer diameter of the examined EEFL was 4 mm with a glass thickness of 0.5 mm, and a total length of 715 mm. The electrode length was 33 mm. The gas pressure and composition were 50 Torr and Ne:Ar with a ratio of 9:1, respectively. By using a square-wave voltage at a frequency of 50 kHz and a duty of 60%, the luminous efficacy η_L was found to be about 52 lm/W. Other parameters were obtained from the measured emitting spectrum as well as the reported phosphor quantum efficiencies.[12] The estimated values were $\eta_E \sim 0.488$, $Q_L \sim 0.8$, $L_e \sim 220\,\mathrm{lm/W}$ at the color coordinates of ($x = 0.26$, $y = 0.23$). The final UV efficiency η_{UV} was obtained from Equation (10.1) to be ~0.58. This value is smaller than the typical value for a hot-cathode fluorescent lamp, which is normally 0.65.[3] This may be due to larger electrode losses for an EEFL.

10.2.4 Lifetime

The lifetime of a fluorescent lamp is conventionally defined by the specific time when the luminance of the lamp is reduced to 50% of its initial value. The luminance decrease occurring during the operation is normally due to phosphor deterioration caused by ion and UV bombardment, in addition to mercury penetration into the glass resulting in decrease of the glass transparency. The lumen maintenance curve of an EEFL is expected to be similar to that of a CCFL because of the similarities of the lamp structure and discharge characteristics. The complete consumption of all the mercury in the lamp terminates the lamp life irrespective of the lumen maintenance. Considering mercury consumption mechanisms, an EEFL is expected to be superior to a CCFL. In the case of a CCFL, internal electrodes are sputtered and eroded continuously due to ions impinging on the electrodes. The sputtered metal atoms combine with mercury atoms to form a mercury–metal amalgam. This is the dominating mechanism of mercury consumption in a CCFL. In contrast, there is no electrode sputtering in an EEFL, which is a favorable condition for reducing the mercury consumption. The amount of consumed mercury in an EEFL was found to be much smaller than that in a CCFL operated at both 25 °C and 0 °C.[13] Moreover, the speed of mercury consumption in an EEFL was barely accelerated even when the lamp temperature was low.

Another issue related to the lifetime of an EEFL is the so-called 'pinhole formation' in the glass tube near or beneath the external electrodes. This refers to the phenomenon of the formation of a small hole in the glass tube when the EEFL is operated at power levels three times or more higher than the normal values.[14,15] Figure 10.3 shows typical shapes of pinholes formed on the electrode surface of an EEFL.[14,15] If the current density at external electrodes increases, the temperature at the electrode becomes higher due to enhanced ionic bombardment and the possible occurrence of corona discharges. These in turn reduce the dielectric breakdown voltage and the resistance of the glass tube, resulting in a further increase of the glass temperature due to energy dissipation in the glass. If the effective voltage applied to the glass becomes higher than the breakdown voltage, or if the local

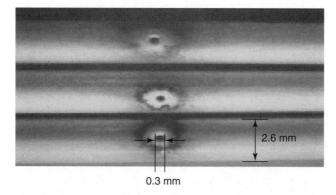

0.3 mm

Figure 10.3 Typical shapes of pinholes formed on the electrode surface of an EEFL (reproduced by permission of the Society for Information Display).[14,15]

temperature of an area of the glass becomes higher than the melting point, a pinhole is formed and the life of the EEFL is terminated. Experiments on various EEFLs have empirically shown that the possibility of pinhole formation becomes very high if the electrode temperature approaches approximately 200 °C. In order to suppress the possibility of pinhole formation, it is important to reduce the current density at the electrodes by increasing the electrode area as well as by maintaining the temperature near the electrodes as low as possible by optimizing the backlight structure for effective heat release from the electrodes.

10.3 Advantages and Disadvantages of EEFL Backlights

10.3.1 Power Consumption

The most attractive feature of EEFL backlight is that multiple EEFLs can be operated with a single inverter, in contrast to CCFLs that are operated individually by independent inverters. In this respect, an EEFL is a more suitable light source for the direct-lit backlight which usually requires a few tens of lamps in a backlight unit. As has been mentioned in the previous section, the parallel driving of the EEFLs is possible owing to the positive resistance characteristic of the lamp voltage–current relationship. The luminous efficacy of the EEFL is comparable to that of the CCFL if the physical parameters of both lamps are identical. However, the power dissipation in the driving circuit becomes lower due to a decrease in the number of circuit elements, in particular, transformers in the driving circuit. Table 10.1

Table 10.1 Comparison of the performances of 26-inch and 32-inch CCFL and EEFL backlights.[16]

Product size (inch)	Power consumption (W)	Luminance on LCD modules (cd/m²)	
		CCFL	EEFL
26	67.2	480	510
32	85.0	450	530
Efficacy comparison (cd/m²/W, %)		100%	106~120%

compares the performances of 26-inch and 32-inch CCFL and EEFL backlights.[16)] The table shows that the power consumption of a 32-inch EEFL backlight is lower compared to that of CCFL backlight of the same size by approximately 15%.

10.3.2 Cost

The reduction in the number of transformers and electric lead wires for an EEFL backlight simplifies the structure of the inverter and the backlight. In addition, the electrode connection can be simplified because EEFLs can be inserted simply into the capping components at both ends instead of being soldered to each lead wire. Figure 10.4 shows an example of the capping structure of the common electrode of an EEFL backlight. Owing to these advantages, not only the power consumption but also the cost of an EEFL backlight can be lower than that of a CCFL backlight. The inverter cost of a 32-inch EEFL backlight is expected to be less than that of a 32-inch CCFL backlight by more than 50%.[17] However, parallel driving technologies of CCFL backlights have also evolved recently, which may make the cost merit of EEFL backlights less significant.

10.3.3 Pinhole Formation

The formation of pinholes at the electrodes may become a serious problem for the EEFL backlight because it suddenly terminates the lamp life during its operation. Even if the possibility of pinhole formation is small, it significantly damages the reliability reputation of EEFL technology. Several improvements have been carried out to prevent the pinhole formation. One is to increase the number of the lamps. For example, for 32-inch TVs, 12–16

Figure 10.4 Capping structure of the common electrode of an EEFL backlight.

lamps are used for CCFLs, while 18–20 lamps are used for EEFLs. This reduces the lamp current and lamp power of each lamp. The other is to increase the capacitance of the external electrodes by making the electrode area larger. This lowers the voltage applied across the glass tube and also reduces the current density of the electrode. A driving method called the 'half-ground driving method' may also be adopted for reducing the electrode temperature further, as will be described in Section 10.4. These modifications widen the drive margin of an EEFL to make the EEFL backlight unit a commercial success.

Table 10.2 summarizes comparisons of detailed specifications and performances of 32-inch CCFL and EEFL backlights. The table shows that the EEFL backlight is superior to the CCFL backlight in several aspects such as the structure of the backlight, the power consumption and the cost.

10.4 Technological Trends of EEFL Backlights

10.4.1 Driving of EEFL Backlights

A half-ground driving method has been developed and adopted for the EEFL backlight. In this method, two synchronized sine (or square) waves,

Table 10.2 Comparison of detailed specifications of 32-inch CCFL and EEFL backlights.

Item		CCFL	EEFL
Structure	Lamp quantity in backlight unit	12~16	18~20
	Transformer quantity in the inverter	12~16	2
	The number of electric wires	24~32	2
Performance	Lamp lifetime (Hr)	~50,000	>60,000
	Lamp voltage (V(rms))*	~1,050	~1,450
	Lamp efficacy	Good	Good
	Backlight efficiency[18]** (lm/W)	45	55
	Power consumption (W)	100~110	~85
Cost[17]	Inverter ($)	33.9	13.6
	Total materials cost ($)	120.7	113.3

*Outer tube diameter = 4 mm.
**On the backlight, without prism film.

180° out of phase with each other, are applied to both electrodes. The applied voltage is divided into these two driving waveforms and thus reduces the voltage amplitude on each electrode and glass wall while keeping the effective lamp voltage across the discharge space unchanged. The driving method is effective for the formation of a uniform plasma column along the tube axis owing to the smaller leakage of the electric field lines through the glass tube.[9] The method also brings about a uniform temperature distribution along the tube axis, resulting in the uniform distribution of mercury during long-term operation. These conditions are important for achieving luminance and color uniformities. The fact of attaining reduced temperatures and voltages also favors the prevention of pinhole formation.

10.4.2 Luminance and Efficacy

The reduction in LC-TV power consumption is a priority, since the low-power requirement is one of the key factors that enables the LC-TV to become more competitive among various flat panel displays. If the luminance of the backlight is insufficient for special applications such as blinking backlight technology which needs very high peak luminance, high-frequency operation in the MHz range might be adopted to obtain a luminance level of $100\,000\,\text{cd}/\text{m}^2$ or even higher.[9] In reducing the power consumption, not only the luminance level but also the luminous efficacy of the lamp is important.

One way of improving the lamp efficacy is to reduce the lamp diameter. If the outer diameters of the EEFLs for a 32-inch backlight are reduced from 4 mm to 3 mm, the luminous efficacy increases from 49 lm/W to 53 lm/W.[19] The tube diameter controls the wall loss of ions and electrons, which in turn affects the electron temperature in the discharge. A reduced diameter increases the ambipolar diffusion loss at the inner wall of the lamp tube, which needs a higher electron temperature to compensate for the loss of the charged particles. A higher electron temperature is expected to give rise to higher efficiency in the generation of UV light for both EEFLs and CCFLs to some degree. However, the lamp impedance increases with a reduction in tube diameter, and thus the lamp voltage increases from 1.4 kV(rms) to 1.7 kV(rms) with the tube diameter changed from 4 mm to 3 mm.[19] This rise in the lamp voltage and current density, however, increases the possibility of the pinhole formation, and thus careful optimization of the electrode structure is indispensable for lowering the lamp voltage and efficient heat release.

Another method for improving the EEFL efficacy is to optimize the driving condition such as the voltage waveform, the duty ratio and the frequency. One example is a 'self-discharge synchronizing operation' of an EEFL driven with square waves.[20] In this operating scheme, the discharge current is designed to change its direction of flow exactly at the time when the applied voltage begins to decrease from its peak. In order to fulfill this condition, it is necessary to synchronize the self-discharge generated by the wall charges to the voltage peak. If the discharge current becomes zero too early or if the current still remains at the timing of the synchronization, the efficacy deteriorates.[20]

Recently, it was found that optimization of the glass properties was effective in improving the efficacy of an EEFL.[21] As has been described in previous sections, the voltage applied to the glass tube and the energy dissipation in the glass material are determined by the dielectric constant and the dielectric loss. Systematic investigations into several glass materials revealed that the EEFL efficiency can be improved by 10–25% by adopting aluminosilicate for the glass tube, as can be seen from Figure 10.5.[21] It was also found that the pinhole reliability of the EEFLs with glass tubes of lower dielectric loss can be improved.

Both a higher dielectric constant and a lower dielectric loss are desirable for improving the performance of EEFLs since the former reduces the voltage applied across the glass tube while the latter is related to the smaller energy dissipation in the glass materials. These conditions are especially favorable when making the lamps longer for larger-size applications.

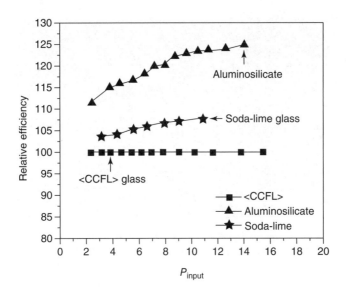

Figure 10.5 Relative efficiency versus input power of EEFLs made by several glass tubes.[21] The efficiency of a CCFL was set to be 100 as a reference (reproduced by permission of the American Institute of Physics).

10.4.3 Bezel Width

One disadvantage of the EEFL backlight is its relatively large bezel compared to that of the CCFL due to the greater electrode length, which is necessary for lowering the lamp voltage and preventing pinhole formation in the EEFL. One of the solutions to this problem is differential diameter EEFL (DEFL) technology.[19] The diameter of the tube at the electrode is larger than the diameter of the main tube where the visible light comes out. Figure 10.6 shows an example of the DEFL. It was shown that both the DEFL and the EEFL exhibit the same electrical and optical characteristics if the electrode areas are the same.[19] This is a very favorable condition for achieving a narrow bezel in the backlight. For example, if the electrode diameter becomes twice that of the main tube, the lamp length can be reduced by approximately 4.6% while other properties remain the same.[19]

10.4.4 Color Gamut

CCFL technologies for obtaining a wide color gamut have recently become popular. Changes of the conventional $Y_2O_3:Eu^{2+}$ (red) and $LaPO_4:Ce,Tb$

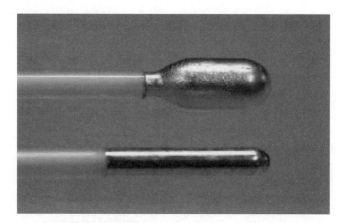

Figure 10.6 Comparison of electrodes of a DEFL with that of a conventional tubular EEFL (reproduced by permission of the Society for Information Display).[19]

(green) phosphors to new phosphors such as $Y(P,V)O_4:Eu^{2+}$ (red), $BaMgAl_{10}O_{17}:Eu^{2+}$, Mn^{2+} (green) and also $Sr_5(PO_4)_3Cl:Eu^{2+}$ (blue) shift the wavelengths of the main emission peaks in addition to removing side peaks in the emission spectrum. As a result, a color gamut of 92% of the National Television System Committee (NTSC) standard has been achieved.[22] The same technique can be applied to the EEFL backlight by using the identical tricolor phosphor combinations. Although the color gamut can be increased with these phosphor materials, there are drawbacks. First, the luminance level of an LCD decreases by 15–20%, because the maximum wavelength of the green emission line shifts away from the peak wavelength of the photopic response curve of the human eye. Secondly, the lumen maintenance curve deteriorates due to the use of BAM-series phosphors for green and/or blue. The lifetime of the phosphor is estimated to be about 30 000 hours, which is approximately 60% that of the normal CCFL/EEFL.

10.5 Development Targets

10.5.1 Long EEFLs for Large LC-TVs

The lamp voltage of an EEFL is higher than that of a CCFL under normal operating conditions because of the additional voltage required by the glass tube. When the lamp voltage exceeds 2 kV(rms), ozone is generated around the electrodes due to corona discharge, which accelerates the electrode oxidation. This is one of the reasons why EEFL backlight technology cannot be applied to large-size LCDs above 50-inch, while CCFL backlight technology is adopted in LC-TVs of all sizes. For an extension of the EEFL backlight unit to larger sizes, it is necessary to improve the lamp performances, in particular, to reduce the lamp voltage below 2 kV(rms).

As mentioned before, the lamp voltage of an EEFL consists of two parts: the voltage applied across the glass tube wall and the discharge voltage. The discharge voltage in the discharge space is again divided into the sheath voltage and the voltage in the positive column. In order to reduce the voltage across the glass tube, it is necessary to increase the capacitance by changing the electrode shape, the glass dimensions and the properties of the glass.[21] For lowering the sheath voltage (i.e. cathode fall), secondary electron emitting materials such as magnesium oxide (MgO) may be deposited on the inner surface of the glass tube under the electrodes. In addition, auxiliary igniting materials like carbon nano-tubes or needle-shaped conducting materials may be dispersed in the lamp or auxiliary electrodes may be designed on the inner or outer surface of the glass tube.

In addition to the problem of higher lamp voltage with increasing tube length, the occurrence of luminance and color non-uniformities along the tube axis is another potential problem for both long EEFLs and long CCFLs. To obtain uniformity, it is important to keep the thickness of the phosphor layer constant and also to keep the distribution of mercury vapor as uniform as possible in the discharge space.

10.5.2 Noncircular Cross Section EEFL

In a conventional direct-lit backlight, a 1–3 mm thick diffuser plate should be put over the parallel tubular CCFLs/EEFLs in order to achieve appropriate luminance uniformity. Some distance between the lamps and the diffuser plate should be maintained for adequate mixing of light from each lamp owing to the relatively large distances between the lamps. Recently, fluorescent lamps having various noncircular cross sections have attracted attention for applications to backlight technology.[23] The cross section of multi-channel flat fluorescent lamps (FFLs) is also noncircular. Owing to the anisotropy of the distribution of light output, there are more degrees of freedom in the arrangement of the noncircular direct-lit EEFL backlights than the cylindrical EEFLs whose distribution of light output is isotropic. There is a possibility that the luminance uniformity of direct-lit backlights can be improved by adopting noncircular EEFLs and by optimizing their arrangement below the diffuser plate.

An external-electrode structure is expected to be more appropriate for noncircular fluorescent lamps because it is simpler to form the electrodes on the noncircular glass tubes by, for example, using a dipping technique. In contrast, it would be more complex to design and optimize hollow cathodes for noncircular CCFLs. The condition for obtaining a high efficacy diffuse positive column in noncircular, low-pressure discharge lamps is sensitive to the dimensional characteristics, in particular, to the aspect ratio or flatness of the noncircular cross section.[23] A concentrated, filament-like discharge of high current density should be avoided to obtain high efficacy.

10.5.3 Mercury-free EEFLs

Recently, there has been a high demand for environmentally favorable light sources. Regulations on hazardous materials described in the Restriction of certain Hazardous Substances (RoHS) tend to prohibit the use of mercury in industrial products. In this respect, it is necessary to develop mercury-free light sources for general lighting and backlight applications.[24] Xenon has been considered as the source element for the generation of UV light owing to its long wavelength among the inert gases. However, the main limitation for wide applications of Xe-type fluorescent lamps is their low efficacy, since the wavelengths generated from a Xe glow discharge are 147 nm and 173 nm, which are shorter than the wavelength emitted from mercury, 254 nm. For example, only a few percent of the input power is transformed into visible light in an aperture-type Xe fluorescent lamp.[10] Intensive studies have been carried out on how to increase the efficacy of Xe discharge lamps during the past decades. For example, it was shown that increasing the Xe pressure was effective in improving efficacy because the relative UV output of 173 nm increased compared to 147 nm[25] resulting in a smaller loss in the phosphor materials during the transformation of the UV photons into visible light.

10.6 Conclusions

An EEFL is a fluorescent lamp operated as a dielectric barrier discharge, based on which the parallel driving of multiple EEFLs with a single inverter can be achieved. The structure of the driving circuit and the backlight, in particular the electrode parts, can be simplified owing to this characteristic. The most outstanding advantages of the EEFL backlight are its low cost, low power consumption, long lifetime, simple lamp structure and simple driving circuit due to the external electrodes. However, EEFL technology has been adopted only in sizes below 42-inch because of the relatively high lamp voltage and the possibility of pinhole formation which increases with increasing tube length.

Further improvements, such as designing noncircular cross sections, the choice of glass material, optimization of the electrode structure and bezel width and obtaining higher uniformities of luminance and color, are in great demand for expanding the application fields of EEFL technology. Basic properties such as plasma parameters,[26] effective capacitance[27] and voltage distribution in the positive column[28] have recently been studied from fundamental aspects, which will accelerate the application of the EEFL backlight technologies to large-size LCDs.

References

[1] Takeda, Y. *et al.* (2003) 'The characteristics of the external-electrode mercury fluorescent lamp for backlight liquid-crystal-television displays', *J. SID*, **11**, pp. 667–673.

[2] Cho, G. *et al.* (2005) 'Glow discharge in the external electrode fluorescent lamp', *IEEE Trans. Plasma Science*, **33**, pp. 1410–1415.

[3] Waymouth, J. F. (1971) *Electric Discharge Lamp*. Place: M.I.T. Press, pp. 24–26.

[4] Raizer, Y. P. (1991) *Gas Discharge Physics*. Place: Springer-Verlag, pp. 381–384.

[5] Lee, S. S. *et al.* (2002) 'Discharge characteristics of external electrode fluorescent lamps (EEFLs) for LCD backlighting application', *SID '02 Digest*, pp. 343–345.

[6] Cho, T. S. *et al.* (2002) 'Effects of electrode length on capacitively coupled external electrode fluorescent lamps', *Jap. J. Appl. Phys.*, **41**, pp. L355–L357.

[7] Cho, T. S. *et al.* (2002) 'Characteristic properties of fluorescent lamps operated using capacitively coupled electrodes', *Jap. J. Appl.Phys.*, **41**, pp. 7518–7521.

[8] Lee, Y.-J. *et al.* (2005) 'Comparative study on sinusoidal and square wave driving methods of EEFL for LCD TV backlight', *Proc. IDW/AD '05*, pp. 1249–1252.

[9] Baba, Y. *et al.* (2001) 'A 100,000-cd/m², capacitively coupled electrodeless discharge backlight with high efficacy for LC TVs', *SID '01 Digest*, pp. 290–293.

[10] Mizojiri, T. *et al.* (2005) 'Energy balance and other characteristics of rare gas fluorescent lamps for image illumination', *J. Light Vis. Env.*, **29**, pp. 104–115.

[11] Choi, J.-Y. *et al.* (2007) 'Light-generating efficiency of external electrode fluorescent lamps (EEFLs) for the backlight applications of liquid crystal display', *Proc. 11th International Symposium on the Science and Technology of Light Sources*, pp. 515–516.

[12] Narita, K. (1985) 'Relative quantum efficiency of various lamp phosphors', *J. Illum. Inst. Jpn.*, **69**, pp. 65–69.

[13] Nishihara, T. (2004) 'Technical trend of light source and inverter for large size LCD TV', *Korea Display Conference Seminar Book*, Volume **2**, pp. 161–180.

[14] Gill, D.-H. *et al.* (2005) 'The lifetime and pinholes in the external electrode fluorescent lamps', *SID '05 Digest*, pp. 1312–1315.

[15] Cho, G. *et al.* (2004) 'Pinhole formation in capacitively coupled external electrode fluorescent lamps', *J. Phys. D: Appl. Phys.*, **37**, pp. 2863–2867.

[16] Han, S.-J. (2006) 'Technology of LCD backlight for TV application', *FPD International 2006 Pre-seminar*.

[17] Displaybank (2006) 'Large-size BLU – analysis and market forecast', *Display Daily*, pp. 69–76 (japan@displaybank.com).

[18] Cho, G. S. (2007) 'Technology of external electrode fluorescent lamps for the backlight of LCD-TVs', *First International Conference of Display for LEDs*.

[19] Kim, J.-B. *et al.* (2006) 'High-performance EEFL backlight system for large-sized LCD-TV', *SID '06 Digest*, pp. 1246–1248.

[20] Cho, G. S. *et al.* (2003) 'Self-discharge synchronizing operations in the external electrode fluorescent multi-lamps backlight', *J. Phys. D: Appl.Phys.*, **36**, pp. 2526–2530.

[21] Cho, G. *et al.* (2007) 'Glass tube of high dielectric constant and low dielectric loss for external electrode fluorescent lamps', *J. Appl. Phys.*, **102**, p. 113307.

[22] Igarashi, T. *et al.* (2005) 'A new CCFL for wide-color-gamut LCD-TV', *Proc. EuroDisplay '05*, pp. 233–235; Kusunoki, T. and Igarashi, T. (2007) 'A high-luminance cold cathode fluorescent lamp for a wide-color-gamut LCD', *IDW '07 Digest*, pp. 2075–2078.

[23] Ko, J.-H. (2005) 'A review on fluorescent lamps having noncircular cross-sections', *IMID '05 Digest*, pp. 1165–1168.

[24] Shiga, T. and Mikoshiba, S. (2002) 'LCD backlights: with or without Hg?' *J. SID*, **10**, pp. 343–346.

[25] Jinno, M. *et al.* (1999) 'Fundamental research on mercury-less fluorescent lamps I, II', *Jap. J.Appl. Phys.*, **38**, pp. 4608–4617.

[26] Cho, G. *et al.* (2008) 'Electron plasma wave propagation in external-electrode fluorescent lamps', *Appl. Phys. Lett.*, **92**, p. 021502.

[27] Lee, T. I. *et al.* (2006) 'I-P relationship and effective capacitance in external electrode fluorescent lamp', *Appl. Phys. Lett.*, **89**, p. 231501.

[28] Choi, J.-Y. *et al.* (2006) 'Determination of the voltage distribution in the external electrode fluorescent lamps for the backlight unit of large-size LCD', *IMID '06 Digest*, pp. 1374–1377.

11

FFL Backlights

G. Kim *GLD Co., Ltd. and Mirae Corporation,* and
S. Lim *Dankook University*

11.1 Introduction

CCFLs (cold cathode fluorescent lamps) and EEFLs (external electrode fluorescent lamps) are mainly used as light sources for LC-TVs. While the LC-TVs
have less limitation in their set thicknesses than laptops and monitors, they
require a higher luminance. Accordingly LC-TVs have to use a large number
of lamps. The narrow tube lamps are located behind the LCD panel. In the
case of CCFLs and EEFLs, 10 or more lamps are required for high luminance
and uniformity. In addition, many related parts and inverters are needed for
these lamps. If, on the other hand, flat fluorescent lamps (FFLs) are used as
light sources for LC-TVs, the number of parts and inverters can be reduced,
providing higher efficiency and lower cost backlight units (BLUs).[1] This is
the reason the development of FFLs is necessary. Figure 11.1 shows a picture
of an FFL in operation.[2]

11.2 The History of FFL Development

Table 11.1 shows the evolution of FFL development. In the early generation,
high luminance was one of the major objectives. In fact early FFLs were made

---

LCD Backlights Edited by Shunsuke Kobayashi, Shigeo Mikoshiba and Sungkyoo Lim
© 2009 John Wiley & Sons, Ltd.

Figure 11.1   Flat fluorescent lamp (FFL).

Table 11.1   FFL development.

Items	Generation 1. (Sand-blasting)	Generation 2. (Glass-forming)	Generation 3. (External electrode)	Generation 4. (Tipless)
Year	~1889	1999–2003	2003–2006	2005–
Method of channel formation	Sand-blasting	Glass-forming	Glass-forming	Glass-forming
Channel type	Serpentine	Serpentine	Straight	Straight
Electrode type	Internal	Internal	External	External
Exhaust tube	Glass tube	Glass tube	Glass tube	Tipless
Maximum size		19" (tiling)	32"	42"
Representative corporation	Thomas electronics[1]	Flat Candle	Samsung Corning	Mirae

mainly as backlights for high luminance LCDs used for avionics and TV applications. A few of the most important objectives included an increase of FFL size and the production of FFLs in volume along with the rapid growth of the LC-TV market. The size of the FFL is important as the sizes of LC-TVs increase. The price of an FFL is also an important issue, so automation of FFL volume production is essential to reduce the production cost.[3]

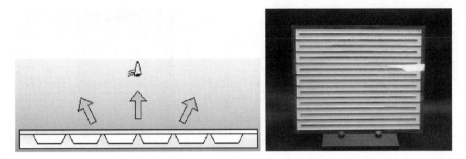

Figure 11.2   FFL channel formed by sand blasting.

11.2.1 Generation 1 FFLs (Sand-blasting)

Figure 11.2 shows the cross section and top view of the channel of a generation 1 FFL. The channels for electric discharges were formed by the sand-blasting technique. To generate stable, low-voltage discharges inside the channels, the depths of the channels should be at least 2–3 mm. The glass plate should then have a thickness of more than 5 mm to form the sand-blast channels. The result of this is that the weight of the FFL increases. In addition, it was difficult to produce such FFLs in volume, so that the production cost of the FFLs was high.

11.2.2 Generation 2 FFLs (Glass-forming)

A glass-forming method was developed in which a special mold was designed to form the channel from a thin glass sheet by heating, and then cooling the mold with the thin glass placed on the mold. Figure 11.3 shows the cross section and top view of an FFL with its channels. Since the substrate can be thin, the weight of the FFL is much lower than that of the sand-blasted FFL.

However, it was not possible to produce FFLs in volume with this method because it took an appreciable time to raise the temperature of the mold in an oven, and it took even longer to cool it down before removing the formed glass. Therefore it became necessary to develop a special glass-forming machine to produce the formed glass at a higher production rate. The size of an FFL has increased as the sizes of LC-TVs increased. Generation 2 FFLs used internal electrodes to drive the lamp, which had a long serpentine channel as can be seen in Figure 11.4. The increase in the FFL size caused an increase in channel length. This resulted in an increase in the firing voltage, which was not desirable for the TV application.

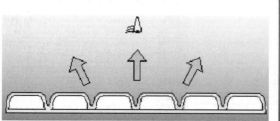

Figure 11.3 FFL channel formed by glass forming and an FFL in operation.

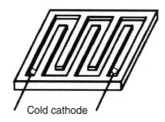

Cold cathode

Figure 11.4 FFL with internal electrode.

11.2.3 Generation 3 FFLs (External Electrode)

It was therefore necessary to find a way of lowering the firing voltage of an FFL. In generation 3 FFLs, a multiple array of channels made by the glass-forming technique was used, as illustrated in Figure 11.5. All the channels were driven using external electrodes with only one inverter. It was found that the structure was effective in lowering the driving voltage whilst maintaining, uniform luminance among the channels. Therefore it became possible to increase the size of an FFL up to 40 inches.[4]

11.2.4 Generation 4 FFLs (Tipless)

Achieving cost reduction in LC-TVs is always the last and biggest issue for the increase of its market share. Therefore, it was necessary to develop an automated TV assembly process for FFL backlights. The number of parts and inverters for the FFL BLU is far fewer than those needed for CCFL and EEFL BLUs, and this helps with automation. Nevertheless, for the automation of a TV assembly, the exhaust glass tip of Figure 11.6 (a) becomes a

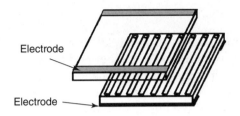

Figure 11.5 FFL with external electrode.

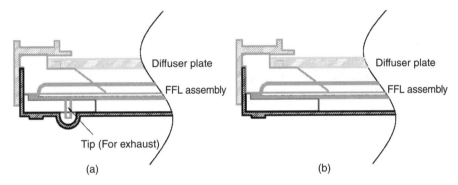

Figure 11.6 (a) FFL with tip and (b) tipless FFL.

handicap. The tip may break during the assembly. The tip also increases the thickness of the BLU. A tipless FFL shown in Figure 11.6 (b) was then developed and assembly automation became easier.

11.3 Characteristics of FFLs

The FFL has the following features:

(a) it simplifies BLU structure and reduces the cost by introducing automated assembly;
(b) it has high luminance and luminance uniformity;
(c) it is more efficient than a CCFL and an EEFL;
(d) it has a long lifetime of 50000–60000 hours;
(e) it is resistant to shock and vibration; and
(f) it does not contain Pb and contains only a minimum amount of Hg.

11.3.1 Structure

An FFL has a formed front glass and flat rear glass on which phosphor is coated. Glass frit is used to seal the upper and lower glasses. External electrodes are provided at both ends of the channels. A gas mixture with a pressure of several tens of Torrs and a fixed quantity of Hg is admitted into the FFL. The cross section of the FFL is shown in Figure 11.7.

11.3.2 Light Emission from FFLs

The light emission mechanism is similar to that for the conventional EEFL as indicated in Figure 11.8. An FFL is a fluorescent lamp which operates in a range of normal glow discharge. A rare gas mixture and a small amount of mercury (a few mg) are introduced into the formed glass which is coated with phosphor. By adding a high electric field between electrodes at both ends of the formed glass, a glow discharge occurs within the channels of the FFL. Excited mercury atoms emit ultraviolet radiation of 254 nm, which excites the phosphor. The visible wavelength depends on the composition of the phosphor materials.

11.3.3 The Fabrication Process

The fabrication process of an FFL is explained in Figure 11.9.

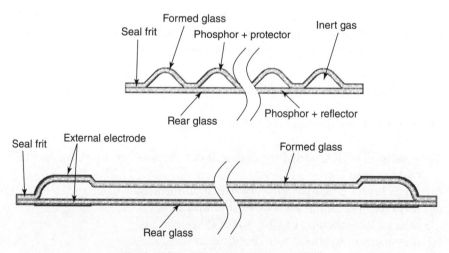

Figure 11.7 Cross section of an FFL.

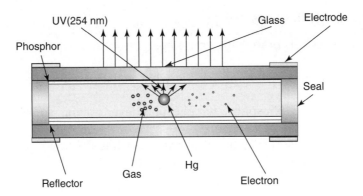

Figure 11.8 Light emission mechanism from an FFL.

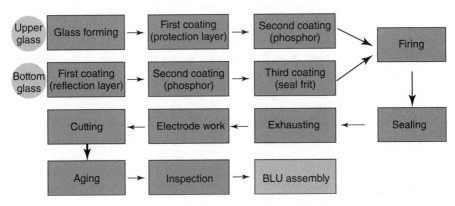

Figure 11.9 FFL fabrication process.

11.3.3.1 Upper Glass

The process consists of the following steps

(a) Glass forming: forming the FFL channel of thin sheet glass with a mold.
(b) First coating (protection layer): the protection layer coating on the glass is to prevent attack from UV radiation.
(c) Second coating (phosphor): the phosphor layer coating is placed over the protection layer.

11.3.3.2 Bottom Glass

The process consists of the following steps.

(a) First coating (reflection layer): the white reflection layer coating is to improve luminance.
(b) Second coating (phosphor): the phosphor layer coating is placed on the reflection layer.
(c) Third coating (seal frit): the glass frit for sealing the upper and bottom glasses is dispensed.

11.3.3.3 Assembly of the Upper and Bottom Glasses

The process consists of the following steps.

(a) Firing: the firing of the protection, reflection, phosphor and frit layers.
(b) Sealing: the sealing of the upper and bottom glasses.
(c) Exhaust: the exhausting of gases and insertion of rare gases and mercury.
(d) Electrode connection: the silver layer coating is applied at the edge of the FFL and electrically connects it to an inverter.
(e) Glass-edge cutting: the edge of the sealed FFL is cut into a proper shape.
(f) Aging: the stabilization of the FFL discharge characteristics.
(g) Inspection: the testing of the external appearance, electronic and optical characteristics.

11.4 Features of the FFL

11.4.1 High Luminous Efficiency and Luminance

Table 11.2 shows the luminous efficiency comparison between various light sources of backlights for LC-TVs. The efficiencies of FFLs, CCFLs and EEFLs are similar because the FFL operation mechanism is almost the same as that of the existing CCFLs and EEFLs.[5] For CCFL/EEFL BLUs, a diffuser plate with high diffusivity is required to eliminate luminance non-uniformity. The transmittance of the diffuser plate is generally low and light is lost. A BLU with an FFL does not need a diffuser plate because of the short distance between the channels. This means that an FFL BLU has higher efficiency and luminance. Table 11.3 shows a luminance comparison of a BLU and LCD with an FFL and with a CCFL with the same power consumption. For the FFL, the optical components consist of three diffuser

Table 11.2 Efficiency of light source for LC-TVs.

Light source	Lamp efficiency (lumens/W)
CCFL/EEFL	~60
FFL	~60
Hg-free Xe lamp	~33
LED	~23

Table 11.3 Comparison of luminance between an FFLBLU and a CCFLBLU.

Items	FFL		CCFL	
Structure	3 diffuser sheets Transparent acrylic plate FFL		3 Diffuser sheets Diffuser plate CCFL Reflector sheet	
Power consumption	103 W	110 W	103 W	110 W
Luminance of BLU	10 300 cd/m²	11 000cd/m²	9100 cd/m²	9600 cd/m²
Luminance of LCD	515 cd/m²	550 cd/m²	455 cd/m²	480 cd/m²
Uniformity	85%	85%	89%	90%

sheets, a transparent acrylic plate and the FFL. For a CCFL, there are three diffuser sheets, a diffuser plate, the CCFL and a reflector sheet.

11.4.2 Long Life

11.4.2.1 Mercury Consumption at the Glass Wall

Figure 11.10 qualitatively describes the correlation between mercury consumption and luminance degradation in fluorescent lamps. Mercury consumption processes take place mainly on the glass wall and on the phosphor particles. The processes are initiated by the mercury ions which bombard the surface of the glass wall during lamp operation.[6] The higher the wall load (lamp power per unit area of glass wall), the more mercury ions reach the wall. The mercury ions then migrate into the glass and are lost from the gas volume. The migration increases with increasing wall temperature. Table 11.4 shows a comparison of the wall load for a CCFL, an EEFL and an FFL. The wall load of an FFL is lower than that of a CCFL or an EEFL. The

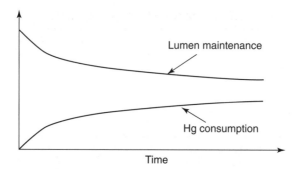

Figure 11.10 Typical behavior of mercury consumption and lumen maintenance in a fluorescent lamp.

Table 11.4 Comparison of wall load.

Light source	Temperature (C)		Barrier coating	Load (W/cm²)	Remark
	Electrode	Center			
CCFL	85	45	Y_2O_3	0.085	5.0 mA
EEFL	87	51	Y_2O_3	0.09	4.0 mA
FFL	60	40	Upper: Y_2O_3 (5 μm)	0.02	Current/channel
			Lower: Al_2O_3 (>5 μm)		3.7 mA

temperature of the wall under the electrode is also lowest in an FFL. There are thin oxide layers and phosphor layers coated on the surface of FFLs. These layers prevent the mercury ions from migrating into the glass wall. For FFLs and EEFLs, there is no mercury consumption due to the formation of an amalgam with the sputtered metal electrode, which is observed in CCFLs.

11.4.2.2 Lifetime

The luminance maintenance curves of FFLs and CCFLs are plotted in Figure 11.11. It is shown that the lifetime of FFLs is superior to that of CCFLs. Figure 11.12 shows the FFL lamp voltage fluctuations measured under normal and elevated temperatures. Figure 11.13 shows the color coordinate fluctuation Δx and Δy of the CIE chromaticity coordinate.

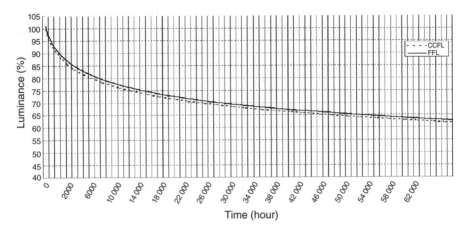

Figure 11.11   FFL lifetime.

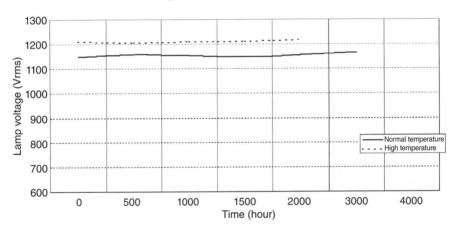

Figure 11.12   Lamp voltage fluctuation.

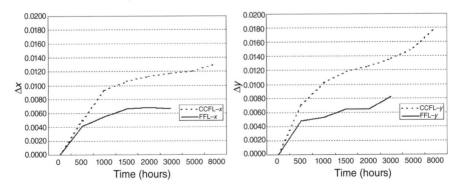

Figure 11.13   FFL color coordinate fluctuation.

Table 11.5 Color-rendering index of various light sources

Light source	*Ra*
General fluorescent lamp	63–77
Three-wave fluorescent lamp	84
CCFL/EEFL	85
FFL	85
Sunlight	100

11.4.3 Wide Color Reproduction

Table 11.5 shows the color rendering indices, *Ra*, of various light sources including a three-wave fluorescent lamp, a CCFL, an EEFL and an FFL. The *Ra* of an FFL is similar to that of a CCFL/EEFL.[7]

References

[1] Anandan, M. and Ketchum, D. (1991) *Proc. SID*, **32**, pp. 137–140.

[2] Hur, J. *et al.* (2000) *SID '00 Digest*, pp. 1033–1036.

[3] Hsieh, D. (2005) *CVCE '05 Digest*, p. 87.

[4] Kim, C. (2005) *Trend and Analysis of BLU Technical Market*, pp. 98–101.

[5] Kim, J. H. *et al.* (2004) *IMID '04 Digest*, pp. 795–798.

[6] Hitzschke, L. *et al.* (2004) *SID '04 Digest*, pp. 1322–1325.

[7] Kumho Electric, Inc. (2003) *2003–2004 Products Guide*, No. 6.

12

Magnetically Coupled Electrodeless Lamps

F. Okamoto

Matsushita Electric Works

12.1 Introduction

Concerns about environmental protection in terms of energy consumption, quantity of materials and the use of hazardous substances are driving various technological improvements for fluorescent lamps. Among them is a magnetically coupled electrodeless lamp. The phenomenon of electrodeless discharge was discovered in 1884 by J. W. Hittorf of Germany. Electrodeless lamps stimulate the gas by using high-frequency electromagnetic induction. The gas then emits ultraviolet radiation which excites a phosphor to obtain visible radiation. Since there are no electrodes – and hence there is no electrode sputtering – lamp life is extended and there is no contamination of the gas.

High-frequency power supplies, however, used to be expensive, so that electrodeless discharge devices tended to be used only for special purposes, for example as light sources for spectrophotometry or plasma generation. Later when high-frequency power supplies became less expensive, the electrodeless lamp was commercialized for lighting in 1990. Several kinds of

LCD Backlights Edited by Shunsuke Kobayashi, Shigeo Mikoshiba and Sungkyoo Lim
© 2009 John Wiley & Sons, Ltd.

high-intensity discharge (HID) lamps were commercialized.[1-8] The operating frequencies of the lamps are restricted to 13.56 MHz (for industrial, scientific and medical uses), as well as 2.65 MHz, 480 kHz, 250 kHz and 135 kHz. The basic operating principles, environmental issues, examples of commercial products, trends in research and development – and also an application as the backlight unit for LCDs – will be discussed in this chapter.[9]

12.2 The Operating Principle of Electrodeless Lamps

In 1947, Babat divided electrodeless lamps into three[10] categories as shown in Figures 12.1(a)–(c).

12.2.1 E Discharge

An AC voltage is applied to electrodes which are attached to the outer surface of the discharge tube (Figure 12.1(a)). An electrostatic coupling generates a discharge. The tube wall acts as a series ballast capacitance of the discharge so that the operating frequency should not be too low. An equivalent circuit for the lamp is shown on the right of the diagram. The lamp is called an external electrode fluorescent lamp (EEFL) as introduced in Chapter 10. Since there are no electrodes in contact with the discharge, the lamp is sometimes referred to as an electrodeless lamp.

12.2.2 H Discharge

The primary coil is wound around the outside of the discharge tube, whereas the secondary 'winding' is formed by the azimuthal current flow in the gas volume (Figure 12.1(b)). The gas atoms are energised by an electromagnetic induction of 1–100 MHz. The discharge is triggered by electrostatic discharges which are caused by a fringing electric field between the adjacent windings. An equivalent circuit diagram is shown on the right of the diagram.

12.2.3 Microwave Discharge

Microwaves in the frequency range 1–100 GHz are generated in a magnetron and led to the cavity through a waveguide (Figure 12.1(c)). In the cavity the microwaves resonate and generate a gas discharge in the tube. The operation is similar to home-use microwave ovens. Since the GHz range frequencies cannot be generated by semiconductors, the system requires a large volume power supply for the lamp.

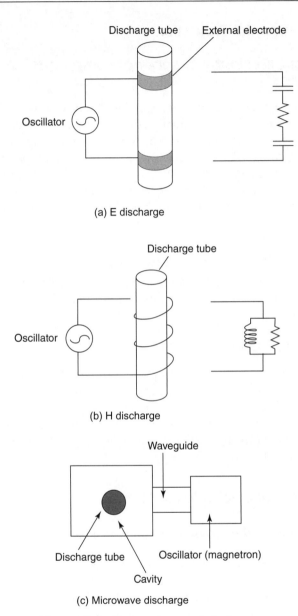

(a) E discharge

(b) H discharge

(c) Microwave discharge

Figure 12.1 Various types of discharges.

12.3 Environmental Protection

12.3.1 Reduction in the Number of Lamp Parts and Materials

Materials which enhance electron emission are often coated on the electrodes of CCFLs. The materials are gradually sputtered by ion bombardments and eventually electron emission is lost. The life of fluorescent lamps for general lighting is determined by electrode life. Electrodeless lamps, on the other hand, have no electrodes, thus assuring a longer life. This means that a smaller number of lamps will be required. If lamp life is extended by five times, for instance, the number of lamps produced can be reduced by a factor of five.

12.3.2 Recycling

The structure of an electrodeless lamp is simple. Also, there is no contamination the of phosphor and glass tubes by sputtered materials, which simplifies the recycling processes.

12.3.3 Energy Saving

Luminous efficacy is higher for lamps with internal electrodes, although frequent on–off switching of these lamps shortens the cathode life. For electrodeless lamps, the on–off action has no influence on lamp life. The annual lamp energy consumption could be reduced if a switching element which senses the human body, for instance, is attached to the lamp for controlling the frequent on–off action of the lamps.

12.4 Features of Electrodeless Lamps

The main features of the electrodeless lamp are given below.

(1) Long life is expected since there is no electrode.
(2) High luminous efficacy is obtained since there is no electrode loss.
(3) Instant ignition and instant re-ignition of the lamp are possible.
(4) Wide variations in the shape of the lamp are allowed.
(5) The build-up of luminous flux on ignition is fast since the volume of the lamp can be made small.
(6) There is no flickering since the lamp is operated at frequencies higher than 100 kHz.

(7) Chemically active gases such as halogens can be introduced into the lamps since there are no electrodes inside the discharge volume. This allows high efficacy and high color rendering.

12.5 Commercial Products with Electrodeless Lamps

Examples of magnetically coupled electrodeless lamps (electrodeless fluorescent lamps) will be explained in this section. Other than the applications for general lighting, these lamps are used by developers of semiconductor patterning, for copy machines, etc.

12.5.1 Outer Coil Lamp

Figure 12.2 shows the first electrodeless lamp in the world, which was commercialized in 1990. A three-turn induction coil is wound around a spherical glass bulb of 35–62 mm diameter. The coil is excited by a 13.56 MHz current.[11,12] The ignition circuit is contained in the lamp base, while the bulb and coil are shielded by a metal mesh to reduce EMI. The lamp is used, for example, for bridge illumination. There are 9 W, 23 W and 50 W lamps[13] whose driving circuit is shown in Figure 12.3. Microwaves generated in an oscillator are amplified and fed to the induction coil through a coaxial cable and matching circuit. A safety circuit limits the operation of the oscillator when, for instance, a sudden change of the load occurs. Figure 12.4 shows a 50 W lamp and its driver.

12.5.2 Inner Coil Lamp

The lamp illustrated in Figure 12.5 has a structure in which a ferrite core with a coil is inserted into a hollow which is made in the glass bulb. The

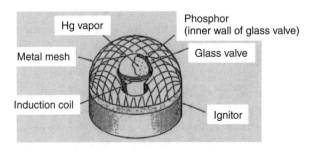

Figure 12.2 Outer coil lamp.

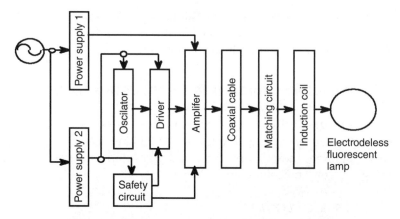

Figure 12.3 Block diagram of driving circuit.

Figure 12.4 Outer coil lamp (13.56 MHz, 50 W) and driver.

ignition circuit of the lamp is contained in a separate metal case and connected to the lamp with a 40-cm long cable. There are 55 W, 85 W and 165 W models of this lamp, all for 2.65 MHz drive and with a life expectancy of 60 000 hours.[14]

Two 135 kHz drive models with 50 W, 4550 lm and 150 W, 13 800 lm were successful in reducing the system power loss and achieving luminous efficacy exceeding 80 lm/W.[15] Figure 12.6 is an external view of the 150 W lamp. The lamp shown in Figure 12.7, which is driven at 2.65 MHz and has an ignition circuit in the lamp base, was commercialized in 1994. Incandescent

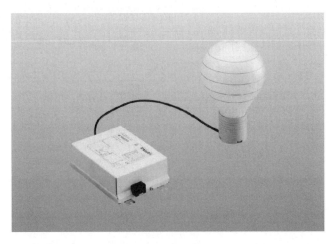

Figure 12.5 Inner coil lamp (2.65 MHz) and driver.

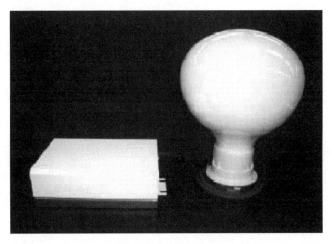

Figure 12.6 Inner coil lamp (135 kHz, 150 W) and ignition circuit.

bulbs, which are widely used, can be replaced by this lamp since it uses an Edison type cap. A transparent conducting film is coated on the inner surface of the bulb to eliminate EMI. The luminous flux output is 1100 lm with a life of 10 000 hours when the power input is 23 W. The lamp shown in Figure 12.8 is also interchangeable with 100 V incandescent bulbs for home use. The discharge volume surrounds the induction coil. There are 12 W and 20 W models with a drive frequency of 48 kHz. The life is 30 000 hours.

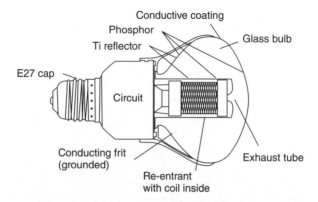

Figure 12.7 Inner coil lamp (2.65 MHz, incandescent light bulb type).

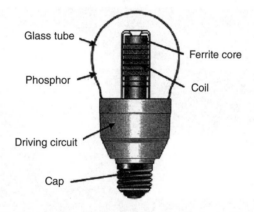

Figure 12.8 Inner coil lamp (480 kHz, incandescent light bulb type).

12.5.3 Toroidal Winding Lamp

A lamp with a toroidal winding, shown in Figure 12.9, was commercialized in 1996. Ferrite cores are attached at both ends of the lamp. The drive frequency is 250 kHz with a power input of 100 W or 150 W. The life is 60 000 hours.[16,17]

12.5.4 Microwave Lamp

A microwave HID electrodeless lamp was first commercialized in 1975.[8] Since then the lamp has been used for printing and photolithography. Halogen compounds, as well as sulfur vapor, are contained in the lamp for light emission. As shown in Figure 12.10, microwaves generated in the magnetron are fed into a metallic cavity resonator to trigger the discharge. The metallic cavity also acts as an EMI shield. For lighting applications, the waveguide is made several tens of meters long, and light is extracted from the waveguide at several points to illuminate rooms.

12.6 Trends in Research and Development

Research and development trends of the magnetically coupled electrodeless lamp are:

(1) extension of life,[18]

(2) obtaining higher output flux as a replacement for the mercury lamp,[19]

(3) containing an ignition circuit in the lamp base as a replacement for the incandescent lamp,[20]

(4) intensity control of the lamp,[21]

(5) luminous efficacy improvement of the overall lamp unit possibly by reducing the operating frequency to 100–500 kHz.[22,23]

12.7 Application to LCD Backlights

Fluorescent lamps and LEDs are used for LCD backlights. In particular, CCFLs are commonly used due to their superior luminance, efficacy, long life, color gamut and reliability. In order to attain high luminance for large-size TVs, a flat backlight which illuminates all the liquid crystal pixels directly from behind the LC panel is desirable. Mercury-free lamps are also being investigated to provide a fast response of luminance.[24] The toroidal lamp of Figure 12.9 operates at a relatively low frequency and its future use as a backlight is expected.

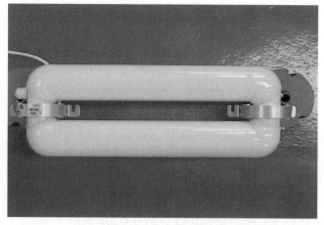

(a) External view of the lamp

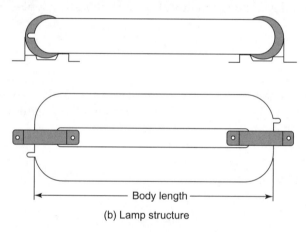

Body length

(b) Lamp structure

Figure 12.9   Toroidal winding lamp (250 kHz).

12.8 Conclusions

Since the first commercialization of the electrodeless lamp there have been many improvements, although the acceptance of these lamps for backlight units has not yet been achieved. Considering the long life of the lamp, fewer environmental concerns and the flexibility of its shape, there should be a demand to adopt electrodeless lamps as the backlight units of LCD-TVs.

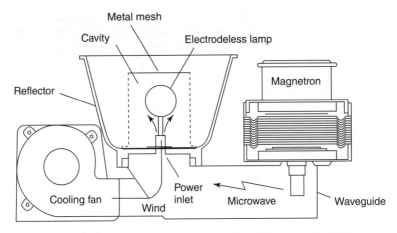

Figure 12.10   Microwave electrodeless HID lamp (2.45 Ghz).

References

[1] Matsushita Electric Works (2001) *'Ever Light' Catalogue.*
[2] Matsushita Electric Works (2004) *Facilities, Stores and Field Illumination Catalogue,* P. T. 127.
[3] Philips Japan (2001) *'QL Lamp System' Catalogue.*
[4] Wharmby, D. O. *et. al.* (1995) 'Low power compact electrodeless lamps', *7th Int. Symp. on Sci. and Tech. of Light Sources,* pp. 27–36.
[5] Mitsubishi Osram (2001) *Lighting Fair 2001 Catalogue.*
[6] (2004)'A fluorescent bulb, "Electrodeless Palook Ball"', *Save Energy, METI,* **56,** p. 50.
[7] Gyoten, M. (2004) 'Electrodeless Fluorescent Lamps', *Japan Electric Lamp Manufacturers Association Report,* No. 461, pp. 78–80.
[8] Dolan, J. T. *et al.* (1992) 'A Novel High Efficacy Microwave Powered Light Source', *6th Int. Symp. on Sci. and Tech. of Light Sources,* pp. 301–302.
[9] Okamoto, F. (2005) 'Present Status and Future Trends of Electrodeless Fluorescent Lamps', *Proc. Illuminating Engineering Institute of Japan,* **89,** pp. 204–207.
[10] Babat, G. I. (1942) *J. IEE, Part* **3,** 94, pp. 27–34.
[11] Yotsumiya, XX *et al.* (1990) *Proc. 1990 Annual Conference of the Illuminating Engineering Institute of Japan.*
[12] Okamoto, F. *et al.* (1992) *Proc. 1992 Tokyo Chapter Conferences of the Illuminating Engineering Institute of Japan.*
[13] Hiramatsu, K. *et al.* (2001) 'Development of Powerful Electrodeless Fluorescent Lamp System', *Matsushita Electric Works Technical Report,* No. 73, pp. 83–86.
[14] (1991) *ILR No. 2,* pp. 44–47.

[15] Kido, H. *et al.* (2005) 'High-Efficiency Low-Frequency Electrodeless Lamp System–EVER LIGHT', *Matsushita Electric Works Technical Report*, 53, pp. 10–15.

[16] Wharmby, D. O. (1995) *7th Int. Symp. on Sci. and Tech. of Light Sources*, No. 2-I.

[17] Shaffer, J. W. and Godyak, V. (1998) 'The Development of Low Frequency, High Output Electrodeless Fluorescent Lamps', *IESNA*, pp. 105–113.

[18] Matsuo, S. *et al.* (2000) 'Analysis on lumen maintenance equation of electrodeless fluorescent lamps', *Proc. 2000 Annual Conference of the Illuminating Engineering Institute of Japan*, p. 57.

[19] Hiramatsu, K. *et al.* (2004) 'High efficacy electrodeless fluorescent lamp system', *Proc. 2004 Annual Conference of the Illuminating Engineering Institute of Japan*, p. 87.

[20] Arakawa, T. *et al.* (2003) 'Discharge properties of the low frequency driven electrodeless fluorescent lamp', *Proc. 2003 Annual Conference of the Illuminating Engineering Institute of Japan*, p. 36.

[21] Masumoto, S. *et al.* (2003) 'A study of dimming method for electrodeless lamp', *Proc. 2003 Annual Conference of the Illuminating Engineering Institute of Japan*, p. 40.

[22] Arakawa, T. *et al.* (2004) 'Buffer gas and driving frequency dependence in a low frequency driven electrodeless compact fluorescent lamp', *Proc. 2004 Annual Conference of the Illuminating Engineering Institute of Japan*, pp. 61–62.

[23] Yamamoto, S. *et al.* (2004) 'Development of electronic ballast for low frequency electrodeless lamp', *Proc. 2004 Annual Conference of the Illuminating Engineering Institute of Japan*, p. 63.

[24] Shiga, T. (2003) 'Mercury-free Xe discharge LCD backlightings', *Proc.2003 Annual Conference of the Illuminating Engineering Institute of Japan*, pp. 251–252.

13

Mercury-free Fluorescent Lamp Backlights

T. Shiga

The University of Electro-Communications

13.1 Introduction

The cold cathode fluorescent lamp (CCFL) which utilizes a mercury discharge is widely used for mid- to large-size LCD backlights. The mercury discharge provides high luminance and high luminous efficacy. The use of mercury, however, causes difficulty in igniting the lamp under low ambient temperatures since the mercury pressure depends strongly on the temperature. In addition, the use of mercury is not recommended for environmental reasons. In order to overcome these problems, mercury-free light sources have been developed. Although LEDs are becoming popular as a small-size LCD backlight source, there are still many issues such as low luminous efficacy, large variation of characteristics, high cost, etc.

Several types of mercury-free fluorescent lamp have been developed. They commonly use Xe discharge and an external electrode structure. The positive column has a tendency to contract to a narrow channel and this tendency is stronger for Xe than Hg. This results in the difficulty of obtaining high luminance and high efficacy.

LCD Backlights   Edited by Shunsuke Kobayashi, Shigeo Mikoshiba and Sungkyoo Lim
© 2009 John Wiley & Sons, Ltd.

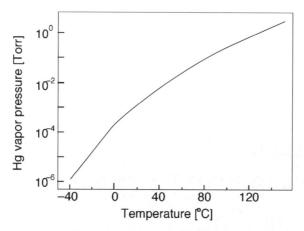

Figure 13.1   Hg vapor pressure versus Temperature.

13.2 Basic Characteristics of Mercury Discharge

A large number of Hg fluorescent lamps are used worldwide. If these lamps were to be destroyed after their lifetime without taking precautions, environmental damage would be substantial. This is one of the disadvantages of the use of mercury discharge. Another disadvantage is that the luminous characteristics strongly depend on temperature. Figure 13.1 shows mercury vapor pressure versus temperature. A change of only a few degrees in the temperature results in one order of magnitude variation of the pressure. This causes difficulty in igniting the lamp at low ambient temperatures. Also, a relatively long build-up time is necessary for the lamp to reach the saturated luminance level. For next generation LCD TV backlights, dynamic backlights with blinking/scanning[1] for improving the moving image quality and local/adaptive dimming[2] for improving the image quality and lowering the power consumption will be introduced. These techniques require fast luminance response and wide dimming capability. The mercury discharge does not satisfy these requirements.

13.3 Basic Characteristics of Xenon Discharge

In the mercury-free fluorescent lamp, a rare gas is used. Among the rare gases (excluding radio-active Rn), Xe is chosen for the following reasons: (1) intense VUV radiation for high luminance, (2) long VUV wavelength with low phosphor damage and high VUV-to-visible conversion efficiency, and

(3) low visible light emission for good color purity. In order to reduce the operating voltage and prolong the lamp life, Ne and/or Ar are added to Xe as buffer gases.

Using Xe as a substitute for Hg causes two difficulties. One is that the Xe positive column tends to contract to form a narrow channel. The contracted discharge has a high current density for which luminous efficacy is low. The luminance is high at the channel, but the total luminous flux is reduced. An effective method to avoid discharge column contraction is to control the discharge so that the current density in the gas volume does not become too high. This can be accomplished by using an external electrode structure, or by optimizing the drive voltage waveform. The basic concept here is to terminate the discharge before neutral atoms are locally heated to create a non-uniform gas density distribution. Figure 13.2 shows a schematic voltage waveform applied across the electrode gap, discharge current and light emission of the external electrode lamp. The space and wall charge distributions at times t_1 and t_2 are also illustrated. Since the electrodes are insulated by dielectric layers, charges produced by the discharge are accumulated on

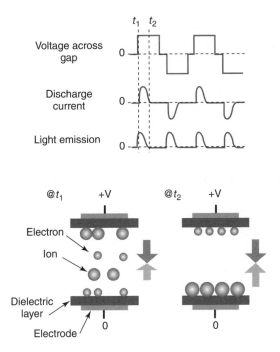

Figure 13.2 Principle of discharge in an external electrode structured lamp.

the dielectric layer. These charges reduce the voltage across the electrode gap, the discharge gradually weakens (at t_1) and eventually terminates (at t_2). An electric field induced by the succeeding voltage pulse has an opposite polarity and therefore the discharge ignites again. In this way, the external electrode structure keeps the AC discharge current low. The structure also results in a longer lamp lifetime because there is no sputtering of metal electrodes.

An other difficulty with substituting Xe for Hg is the degradation of the luminous efficacy due to self-absorption and de-excitation phenomena. Xe fluorescent lamps utilize VUV radiation which consists of 147 nm resonance radiation and a continuum with a center wavelength at 172 nm. The resonance radiation is likely to be absorbed by the neighboring ground state Xe atom, which is called imprisonment of radiation. The effective lifetime of the excited Xe is then prolonged. Thus the excited Xe atoms are more likely to experience collisions with free electrons, leading to the destruction of the excited state (de-excitation). As the discharge current density is increased, there is a higher probability that destructive collisions will occur. This causes saturation of the luminance and a decrease in the luminous efficacy along with an increase in discharge current. In contrast, there is no such phenomenon for the continuum radiation. The radiation comes from Xe excimers that are mainly created by three body collisions of Xe and its excited state. To suppress the reduction of the efficacy due to self-absorption and de-excitation phenomena, formation of a low current density discharge and utilization of excimers by increasing Xe pressure are effective. It should be noted that there is a drawback that the operating voltage becomes high at elevated pressures.

13.4 Mercury-free Xe Discharge Fluorescent Lamps

The discharges utilized in mercury-free lamps are classified into three types: cylindrical discharge, micro-discharge and flat discharge.

13.4.1 Cylindrical Discharge Type

The structure of a cylindrical discharge tube is basically identical to that of Hg-CCFLs, with Hg replaced by Xe. To avoid the discharge contraction, an Xe CCFL with a new structure was developed (Figure 13.3).[3] An inner electrode is positioned at one of the edges of the glass tube, while a wire electrode is coiled around the outer surface of the lamp. The structure,

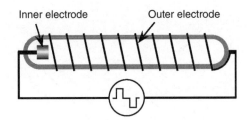

Figure 13.3 Cylindrical discharge type lamp.

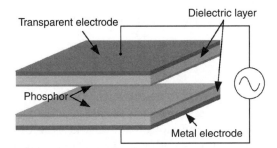

Figure 13.4 Micro-discharge type lamp.

together with a 20 kHz drive, provides a cylindrical discharge without contraction. To obtain uniform axial luminance distribution, spacing of the wire electrode may be varied along the tube axis.

13.4.2 Micro-discharge Type

There are two types of micro-discharge lamps. One of them utilizes the micro-discharges formed between parallel electrodes which are covered with dielectric layers.[4–6] Figure 13.4 shows the basic structure of the lamp. The spacing of the two plates is 0.5–2 mm. Pure Xe, or a Xe–Ne mixture, is contained at a pressure of 10–80 kPa. Phosphor is coated on the dielectric layers. When the lamp is driven with a 20–30 kHz a.c. voltage of 1000 Vrms, many micro-discharges with a diameter of about 0.1 mm are formed between the electrodes. By controlling the input power, the micro-discharges spread uniformly over the entire volume, and a uniform emission is obtained from the lamp. Luminance of 3500 cd/m^2 with luminous efficacy of 27 lm/W were attained in a 3.5-inch diagonal lamp.[4] When the lamp size is enlarged, the discharge characteristics are almost the same as long as the gap and input power per unit area are kept constant. When the lamp size becomes large,

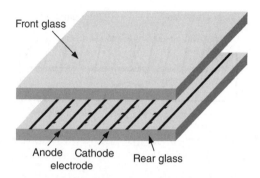

Front glass

Anode   Cathode   Rear glass
    electrode

Figure 13.5 Triangle shaped micro-discharge type lamp.

spacers are necessary in order to maintain constant plate separation. A diffuser plate should then be employed to eliminate the shadows of the spacers.

Figure 13.5 shows the structure of the second type of micro-discharge
lamp.[7–9] Anodes and cathodes with projections are formed on the rear glass
plate. The distance between anodes and cathodes is about 10 mm. The electrodes are covered with a dielectric layer. Front and rear glasses are sealed
with a gap of a few millimeters and Xe gas is filled at a pressure of 13 kPa.
When a pulse voltage is applied between the electrodes, many triangular
micro-discharges are formed between them. A diffuser plate is used to eliminate the shape of the discharges. This lamp can also be made larger without
change of discharge characteristics if the discharge gap and gas pressure are
kept constant. A 32-inch diagonal micro-discharge lamp having an improved
front glass plate shape was developed.[10] Luminance of 6000 cd/m² was
obtained when the input power was 185 W. The lifetime is the same as that
of a conventional CCFL backlight.

13.5 Mercury-free Xe Flat Discharge Lamp

13.5.1 Flat Discharge

A discharge can be generated in a rectangular volume in which a uniform
positive column spreads over the entire volume. Generally the positive
column has a strong tendency to form a narrow channel in a wide discharge
space. To avoid contraction, the gap between the front and rear glass plates,
the electrode configuration and the driving condition should be optimized.

A uniform positive column provides high efficacy and high luminance uniformity of over 90%. A simulation study shows that the UV production efficiency is about 80%.[11]

13.5.2 Structure and Driving of Lamp

Xe flat lamps with diagonal sizes ranging from 0.5-inch to 5.2-inch have been developed[12] for LCD backlights. Also a lamp with an active area 444 mm × 27 mm has been developed for backlights and general lighting.[13] The structure of the lamp is shown in Figure 13.6. The lamp has a simple structure with a front glass plate and a rear glass vessel. On the inner side of the front glass plate are two thick-film Ag electrodes which are covered with a 0.06-mm thick transparent dielectric layer. Tri-color-phosphor is deposited on both the front and rear glass except for the regions above the electrodes. When the lamp is used for backlighting, the front phosphor layer is made thinner than the rear. The lamp contains a mixture of Ne + Ar + Xe gases at a total pressure of 10 kPa. Vacuum ultraviolet radia-

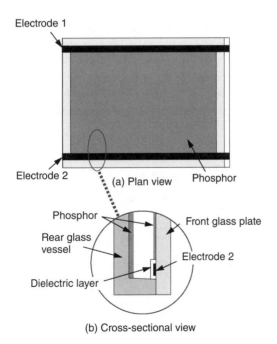

(a) Plan view

(b) Cross-sectional view

Figure 13.6 Structure of flat discharge lamp.

tion of wavelength 147 nm from Xe atoms excites the phosphor. The use of MgO, which has a relatively high secondary electron emission coefficient, on the dielectric layer improves the luminous characteristics, lifetime and dark-room ignition.

The two electrodes are driven with trains of 10–100 kHz rectangular pulses. The two pulses have an identical amplitude V, pulse width t, and pulse interval T, but are 180° out of phase. When the values of V, t and T are properly chosen, a flat discharge can be obtained. If the discharge area contracts, the current density at the electrodes increases and sometimes destroys the insulation of the dielectric layer.

13.5.3 Comparison of Hg and Xe Discharges

Figure 13.7 compares the luminance and efficacy of the 5.2-inch diagonal Ar–Kr–Hg and 2.8-inch diagonal Ar–Ne–Xe lamps. These lamps have similar electrode structures. The peak luminance of the Xe lamp is half that of the Hg lamp. The peak efficacy of the Xe lamp is two-fifths that of the Hg lamp. The efficacy of the Xe lamp exceeds that of the Hg lamp at low luminance levels since the lamp temperature is too low for the Hg discharge to operate.

A luminance of 11 000 cd/m^2 and luminous flux of 860 lm were attained with the previously introduced 444 mm × 27 mm lamp. The luminance and luminous flux are comparable to that of a commercially available FL15 (a 15 W Hg fluorescent lamp). The efficacy, however, was halved. Improvement of the efficacy was attempted by increasing the Xe partial pressure.[14,15] Because contraction reducing becomes more likely with increasing Xe pressure, this was overcome by reducing the electrode width. Also a MgO layer

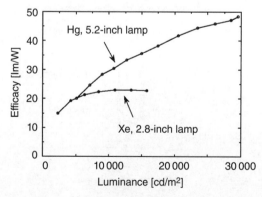

Figure 13.7 Efficacy versus luminance of Xe and Hg flat lamps.

was provided. This resulted in an efficacy of 35 lm/W with luminance of $10000 \, \mathrm{cd/m^2}$ being obtained with a 2.8-inch diagonal lamp.

13.6 Conclusions

Mercury is frequently used in fluorescent lamps for LCD backlighting. The use of mercury, however, is being discouraged due to environmental protection issues. Also the fact that Hg vapor pressure depends on the device temperature limits the adoption of these lamps for various uses. To solve these problems, mercury-free Xe discharge fluorescent lamps were developed. One of the most important factors for obtaining the best optical performances of the xenon fluorescent lamp is to diffuse the discharge without contraction. The luminance and luminous flux of a Xe lamp are comparable to those of an Hg lamp, but luminous efficacy is approximately halved.

References

[1] Fisekovic, N. *et al.* (2001) 'Improved Motion-Picture Quality of AM-LCDs Using Scanning Backlight', *Asia Display/IDW '01*, pp. 1637–1640.

[2] Shiga, T. and Mikoshiba, S. (2003) 'Reduction of LCTV Backlight Power and Enhancement of Gray Scale Capability by Using an Adaptive Dimming Technique', *SID '03 Digest*, pp. 1364–1367.

[3] Noguchi, H. and Yano, H. (2000) 'A mercury-Free Cold Cathode Fluorescent Lamp for LCD Backlighting', *SID '00 Digest*, pp. 935–937.

[4] Urakabe, T. *et al.* (1996) 'A Flat Fluorescent Lamp with Xe Dielectric Barrier Discharges', *J. Light Vis. Env.*, **20**, pp. 20–25.

[5] Kwak, M.-G. *et al.* (1998) '2.3" Flat Fluorescent Lamp with Full Area Electrodes and Hg-Gas Free', *18th Intl Display Res. Conf.*, pp. 467–470.

[6] Choi, J. Y. *et al.* (2000) 'Mercury-Free 18" Class Flat Fluorescent Lamp with Good Uniformity', *IDMC '00*, pp. 231–232.

[7] Ilmer, M. *et al.* (1999) 'Efficacy Enhancement of Hg-Free Fluorescent PLANON Backlights by Controlled Atmosphere Fusing (CAF)', *IDW '99*, pp. 1107–1108.

[8] Ilmer, M. *et al.* (2000) 'Hg-free Flat Panel Light Source PLANON – a Promising Candidate for Future LCD Backlights', *SID '00 Digest*, pp. 931–933.

[9] Ting, C.-C. *et al.* (2005) 'The High Efficiency and Long Life Span Plasma Flat Fluorescent Lamp Equipped with Innovative Dual Driving Waveforms', *IDW/AD '05*, pp. 1265–1268.

[10] Hitzschke, L. *et al.* (2004) 'A 32-in. Integrated Hg-free Lamp that Eliminates Problems of Backlights with Multiple Lamps', *SID '04 Digest*, pp. 1322–1325.

[11] Shiga, T. *et al.* (2003) 'Study of Efficacy in a Mercury-Free Flat Discharge Fluorescent Lamp using a Zero-Dimensional Positive Column Model', *J. Phys. D: Appl. Phys.*, **36**, pp. 512–521.

[12] Ikeda, Y. *et al.* (2000) 'Mercury-Free, Simple-Structured Flat Discharge LCD Backlights Ranging from 0.5 to 5.2-in. Diagonals', *SID '00 Digest*, pp. 938–941.

[13] Shiga, T. *et al.* (2001) 'Mercury-Free Xe Flat Discharge Lamps for Lighting', *J. Light and Vis. Env.*, **25**, pp. 10–15.

[14] Shiga, T. *et al.* (2004) 'Efficacy Improvement of Mercury-Free Flat Discharge Lamps by Increasing Xe Partial Pressure', *J. Illuminating Eng. Inst of Japan*, **88**, pp. 517–521 (in Japanese).

[15] Lee, J. K. *et al.* (2005) 'High Efficiency Mercury-Free Flat Light Source for LCD Backlighting', *SID '05 Digest*, pp. 1309–1311.

14

LED Backlights

M. Zeiler and J. Hüttner

OSRAM Opto Semiconductors GmbH

14.1 Introduction

Over the past three decades, light-emitting diodes (LEDs) have been used increasingly for a wide variety of industrial applications, ranging from consumer electronics to traffic signs and numerous other uses. As the technology has improved in recent years, the use of LEDs has especially grown in more sophisticated applications such as LCD panels for products both large and small. In particular, LED market growth is being driven by their ability to satisfy manufacturers' desire for thin, high-brightness displays possessing a rich color gamut, regardless of the application size. As a result, phosphor-converted white LEDs are emerging as a technology-of-choice for small- and medium-size LCD displays because they ideally meet the need for miniaturization in mobile devices such as cellular telephones. Further, because of the inherent brightness levels now found in white and single-color LEDs, they are also becoming more popular for the backlighting of larger LCD displays, including monitors and flat-screen televisions which leverage the homogenous and uniform illumination of red–green–blue (RGB) LEDs. One of the chief advantages of LEDs in this application area versus competitive backlighting technologies is their color gamut of more than 110% of NTSC.

LCD Backlights   Edited by Shunsuke Kobayashi, Shigeo Mikoshiba and Sungkyoo Lim
© 2009 John Wiley & Sons, Ltd.

The inevitable conclusion to draw from all of this is that improvements in LCD backlighting will become increasingly predicated on LED technology enhancements – meaning that LEDs will play a meaningful role in the manufacturing of some of the most exciting, fastest-growing consumer technologies in the world. This would especially seem to be the case when, upon even deeper inspection, it becomes apparent that LEDs possess many performance and other advantages compared with conventional lighting technologies. In fact, at this stage, the greatest challenges to their increased market penetration are more price-related than performance-related – a consideration that this chapter will explore in more detail later. However, as volume purchasing increases – a strong likelihood – pricing should eventually scale back accordingly. The balance of this chapter discusses the make-up of LEDs, their benefits and disadvantages compared with conventional lighting technologies, and their use in a variety of critical LCD applications.

14.2 LED Device Principle

While discussing the future of LED technology and specifications, it is instructive to understand their basic construction. As illustrated in Figure 14.1, an LED consists of several layers of semiconducting material. When an LED is subjected to a DC voltage, light is generated in the active layer; the generated light is radiated directly, or by reflection.

In contrast to conventional lamps, which emit a continuous spectrum, an LED emits light of a specific color, depending on the material that is used. There are in fact two systems of materials – AlGaInP and InGaN – that are

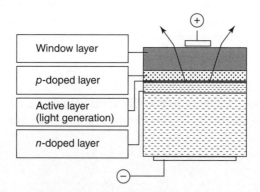

Figure 14.1 Cross section of an LED.

used to produce LEDs with high luminance in all colors ranging from blue to red, as well as white (by luminescence conversion). Depending on the system type, different voltages are required in order to operate the diode in the conducting direction.

14.2.1 Benefits of LED-based Backlight Solutions

Compared with conventional cold cathode fluorescent lamps (CCFLs), LED backlight solutions offer LCD manufacturers a wide range of performance benefits. Above all, LEDs produce a far superior image in terms of color gamut and saturation. LEDs also offer greater contrast and brilliance (they can be switched in less than 100 ns), and unlike CCFL light sources, contain no mercury – making them more environmentally friendly. LEDs are also vibration and shock-proof, and have exceptionally long service lives. Contingent on the application, environment, driving considerations and other factors, they operate for 50 000 hours or longer, over a temperature range of –40 °C to 85 °C.

Generally speaking, LED backlight solutions possess great flexibility, and as such they can be extended to any diagonal size. Further, if a consistent LED packing density (LEDs per unit area) is maintained, then the luminance remains constant for all backlight unit (BLU) sizes. As a consequence, adapting the luminance to specific requirements necessitates alterations to either the LED packing density or the drive current.

14.2.2 Requirements for BLU Systems

The considerable and wide-ranging performance benefits of LEDs in turn have a direct influence on their design specification for backlight applications. The luminance level on an LCD screen should typically be between 200 cd/m² for monitor applications and 500 cd/m² for large-area television applications, with a brightness uniformity of better than 85% . The wavelength ranges of red, green and blue LED dies need to be specified in order to achieve a color gamut greater than 100% NTSC. The mixing ratio of the luminous flux needs to be chosen for color temperature ranges between 6500 K and 12 500 K, depending on factors such as specifications and customer preferences, which vary by region.

For the backlight of a 32-inch LCD panel, power consumption should be well below 150 W, which is comparable to the power consumption of a conventional backlight based on CCFL technology. The key is to keep the number of LEDs to a minimum, which helps ensure competitive cost levels and high-quality backlights with very low levels of failures in time (FIT).

Special long-life housing materials and robust designs for the packages and dies have to be taken into consideration in order to achieve a lifetime of more than 50 000 hours, which is comparable to, or better than, CCFL lifetime. The essential point to bear in mind is that different LED arrangements and backlight technologies have to be developed in order to adapt to BLU sizes as well as customer requirements such as power consumption, brightness requirements, BLU thickness, uniformity requirements, available space, preferred drive conditions, requirements for BLU cost and thermal design capabilities.

14.2.3 Two Generic Designs of LED Components

In principle, there are two different ways in which LED components can be arranged (Figure 14.2). The so-called 'light-guide arrangement' or 'indirect backlight' generates very thin (from 1–10 mm) and uniform backlight units. On the other hand, the 'direct backlight arrangement' offers low power consumption, good thermal design and excellent scalability at a thickness of between 25 mm and 40 mm. The light-guide arrangement can be deployed, with good optical efficiency, for backlights up to approximately a 20-inch diagonal, while the direct backlight arrangement is essentially unlimited in terms of diagonal. Both technologies can be adapted to different specifications and requirements.

Given the overall nature of LED backlights using multiple primaries with different opto-semiconductor devices, a color and brightness feedback system is most highly recommended. This is because LEDs have different thermal and aging behaviors for different wavelengths. To that end, a sensor

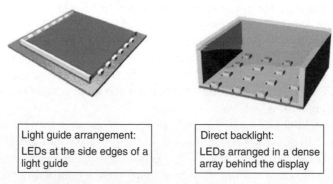

| Light guide arrangement: | Direct backlight: |
| LEDs at the side edges of a light guide | LEDs arranged in a dense array behind the display |

Figure 14.2 Basic LED backlight technologies.

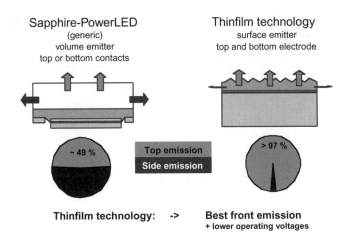

Figure 14.3 Top/side emission ratio of a Thinfilm chip (run on)(right) compared to a conventional Sapphire chip (left).

feedback loop is needed in order to ensure stable white point settings over the operating life.

14.2.4 Thinfilm and ThinGaN® Technology

A new generation of chips – 'Thinfilm' AlGaInP chips for red light and 'ThinGaN®' for blue and green lights – were initially developed to enhance LED current and brightness. Because of their design (shown in Figure 14.3) these chips are pure top surface emitters, with a true Lambertian radiation pattern allowing them to offer all of the benefits associated with high efficiency and outstanding power levels. Unless otherwise noted, henceforth in this chapter both 'Thinfilm' AlGaInP and 'ThinGaN®' chips will be referred to as 'Thinfilm' chips. The Thinfilm chip comprises a substrateless thin active layer centered between a rough surface on the top and a highly reflective metal mirror on the bottom, which is attached to the carrier substrate.[1–4]

14.2.5 Multiple Benefits

The unique design of Thinfilm chips increases efficiency in many ways. First, their efficiency is increased (and forward voltage is decreased) because of the chips' thin active layer extraction. Further, Thinfilm chips are pure top surface emitters (>97%), and the related radiation characteristics show a true Lambertian emission pattern. Consequently, the reflector plays a less pivotal role in Thinfilm chips than is the case in conventional volume emitting chips.

Top surface emission enables excellent optical designs to be implemented, such as for light-spreading optics in LED backlight systems, as stray light can be minimized and hot spots avoided. As an additional consequence of pure surface emission, luminous flux scales up with the size of the chip. This makes it possible to implement very large chips for high flux operation at high currents. On the other hand, conventional volume emitting chips with transparent substrates do not scale up in the same ratio, and as a result lose efficiency as their size increases.

14.3 LED Backlight Solutions for Different LCD Sizes

14.3.1 LED Backlighting of LCD Desktop Monitors

LED brightness has consistently increased in recent years, creating opportunities in a variety of new market segments and application areas. RGB SMD LEDs have become particularly prevalent in the LCD display market because they offer several advantages over existing CCFL technology for use in backlight units. Among these advantages are the exceptional image quality, which results from an increased color gamut and the compensation of blurring effects by means of pulsed LEDs. Furthermore, the color point in these kinds of LEDs can be set to any location within the color triangle, and can be color stabilized and dimmed with the appropriate LED driver circuitry. In addition, LEDs provide a high level of reliability – as mentioned, a typical lifetime exceeds 50 000 hours – and are free of potentially hazardous materials such as mercury and lead. If faster-switching LCDs become available in the future, it is quite likely that the color filter in LCD displays will be eliminated entirely and the color display could be achieved directly by means of 'sequential color'.

The six-lead MULTILED® LRTB G6SG of Figure 14.4 is a multi-chip SMD LED which contains three chips for the colors red (R), true-green (G) and blue (B), using Thinfilm-based technology. This LED is characterized by the high optical efficiency of the chips used and the earlier-referenced extended lifetime resulting from silicon packaging, which allows the LED to be used in demanding backlighting applications.

14.3.2 Example of an Indirect LED Backlighting for a 19-inch LCD Monitor

It is instructive to review the performance of a commercially available, 19-inch TFT monitor which was modified with LED backlighting. Only the light source was modified. The two CCFLs located on the long sides of the light

Figure 14.4 Six-lead MULTILED®.

guide were removed and were each replaced with a strip of insulated metal substrate (IMS) PCB containing 77 LEDs. For the entire backlight, 154 LEDs were used, with a pitch of 5 mm as shown in Figures 14.5 and 14.6. Because of this light-guide approach, it was possible to make the backlight very thin (approximately 10 mm). LED primary colors vary in intensity, depending on the desired color coordinates for the RGB mixture. As a result, RGB ratios need to be adjusted based on the type of LEDs used as well as their brightness and wavelength groups.

14.3.3 The Results

The power consumption per LED was around 190 mW (30 W for all LEDs), and provided a luminous flux of approximately 5.5 lm for the backlight, leading to 250 cd/m² at the LCD. For this specific modification, the PCB strips were mounted in the existing housing. In order to guarantee the stable continuous operation of the LEDs, thermal dissipation must also be taken into account. In this case, it was sufficient to achieve passive cooling by means of ventilation slots in the housing and by mounting the IMS PCBs on thin heat sinks. Appropriate series resistors were used to limit the maximum current to the LEDs; fine adjustment of the RGB ratios was achieved by means of pulse-width modulation.

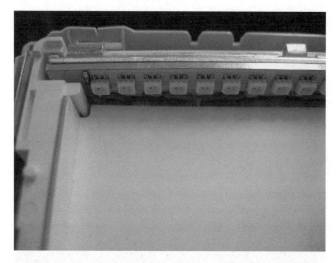

Figure 14.5 Section of the LED strip which replaces the CCFL tubes.

Figure 14.6 Upper left-hand corner of the LED backlight in operation.

14.3.4 LED Backlighting of Notebook LCDs

Notebook users are the most demanding in terms of their expectation for long battery life; few things are more frustrating to a business person, for example, than losing their power supply in the midst of a long-distance flight. Currently, the display subsystem accounts for the highest proportion of the power consumed inside a notebook (approximately 30%). While the LCD itself only requires a small fraction of that power, the backlight is responsible for the relatively high power dissipation.

This once again underscores the potential of modern and efficient LED technology, which can replace conventional backlighting and lead to a drastic reduction in power consumption. Because requirements concerning color reproduction are fairly minimal, white LEDs can be used in this scenario in place of RGB backlighting. The white Micro SIDELED® is highly suitable for this application.

14.3.5 Case Study: Indirect LED Backlighting for a 15.4-inch Notebook Display

For the following test, two identical 15.4-inch notebooks were used. One was modified with LED backlighting, while the other was not. An overview of the components is shown in Figure 14.7. The picture shows the LCD and the backlight unit as separate entities. The backlight itself comprises the CCFL and the light guide with its associated optical films (diffuser sheet, brightness enhancement films and other parts.) The retrofit mainly consisted of replacing the CCFL with a PCB strip with a linear LED arrangement. An external power supply was used, but it is recommended that the existing onboard voltages in the notebook be used by integrating appropriate LED drivers.

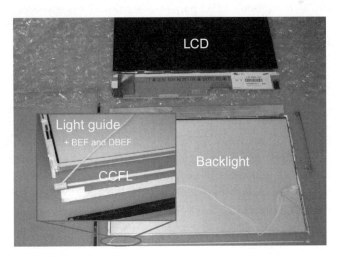

Figure 14.7 Overview of the components.

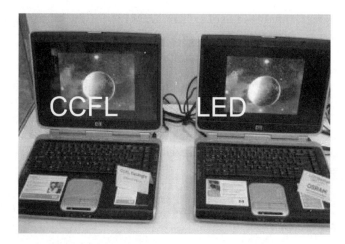

Figure 14.8 Display with conventional CCFL backlighting (left) and with LED backlighting (right).

When compared side by side, it is difficult to distinguish between the CCFL and LED versions, as shown in Figure 14.8, an observation that is supported by the tangible measurement results. For both displays, the same maximum luminance of approximately $180\,cd/m^2$ was achieved, with a uniformity of approximately 80%. The color gamut for both versions was about 40% of NTSC (in notebook LCD panels, the color filters are broader than in desktop panels, which leads to a higher transmission at the cost of a smaller color gamut).

The key finding, however, came when the relative power consumption of each notebook was compared. In Figure 14.9, the power consumption of the LED backlight is represented by the curve for 'LCD A', whereas 'LCD B' represents the CCFL version. The significantly lower power consumption of LED backlighting at all luminance levels is obvious. For example, at $60\,cd/m^2$, where the CCFL consumes 3.37 W, the LED power consumption is below 1.5 W. The notebook user can benefit directly from these power savings, resulting in a greatly extended battery life. LED efficiency has not yet reached its peak, so it should be possible to reduce the backlight power consumption in the near future.

14.3.6 Technology for Large-screen TV Applications

LED technology is also playing an increasingly vital role in the manufacture of large-screen televisions. The especially suited Golden DRAGON® ARGUS®

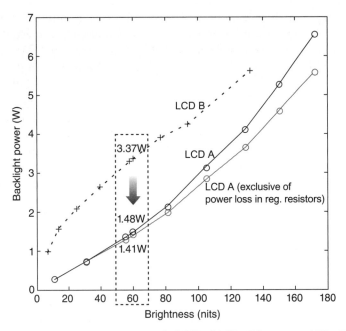

Figure 14.9 Power consumption of CCFL ('LCD B') versus LED ('LCD A') backlighting.

LED is based on the available high-power package Golden DRAGON® LED combined with a widely radiating top-emitting lens.[5] The lens deflects the emitted light in such a manner that emission from the point light source is transformed into a flat and uniform distribution of light. A direct backlight design based on this LED/lens combination results in highly efficient (>85%) LED backlight units.

14.3.7 System Setup

For the construction of white LED backlighting with RGB light sources, different colored LEDs are mixed to produce a uniform white color. This mixed color can be adjusted according to the intensities of the individual primary colors. The mixing ratio is dependent on the wavelength and purity of the primary colors and varies over different LED groups. In principle, the number of primary colors used can exceed three and achieve an even higher color gamut and better color reproduction. Nevertheless, in the following discussion only three primary colors will be considered.

The advantage of the wide radiation characteristic of the Golden DRAGON® ARGUS® is that the primary colors overlap over a very large area and a uniform color mixture can be achieved within a minimum depth. For optimum color uniformity, it is recommended that the individual colored LEDs should be organized in a compact cluster arrangement.[6,7] When defining clusters, the mixing ratio for the required white point should be taken into account. A typical white point of $x = 0.33$ and $y = 0.33$ (neutral white) requires a significantly higher percentage of green than red or blue luminous flux. Therefore, it makes sense to add an additional green LED to form an RGGB cluster, especially since the green LEDs do not then need to be over-driven and effects related to color shifting, aging and reduction of efficiency can be minimized.

If more than three primary colors are used, the number of LEDs forming a cluster should be increased. Depending on the white point required or the LED brightness, various color combinations can be employed. For the design of a complete backlight unit using three primary colors, several RGGB clusters are systematically arranged in an equally spaced array.

In test constructions, typical depths for the backlight of 35–50 mm could be achieved, depending on the uniformity requirements. Figure 14.10 shows a cross-sectional schematic of the basic construction of LED backlight units[8] with Golden DRAGON® ARGUS®. The inner areas, except the upper surface, should be covered with reflective foil to let the light reflect in multiple ways. The distance between the different clusters depends on the overall brightness requirements. On the top side, a set of foils containing diffusers, BEF and DBEF should be placed, depending on the optical requirements. On the back, the heat can be distributed by using a metal-core PCB (MCPCB). Depending on the specification, there is no further need for active cooling.

In the upper left-hand corner of Figure 14.11, for example, a hexagonal cluster arrangement is shown. A constant distance between LEDs allows the smallest possible display depth to be achieved. Edge effects may be noticeable, however, since the density of light at the edges is smaller, due

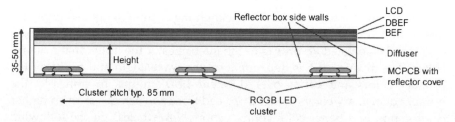

Figure 14.10 Cross section of a typical LED backlight unit stack with Golden DRAGON® ARGUS®components.

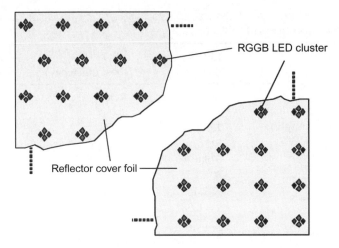

Figure 14.11   Two-dimensional arrangement of RGGB LED clusters.

to this arrangement. In the second example pattern, shown in the bottom right-hand corner of Figure 14.11, the rectangular arrangement of the Golden DRAGON® ARGUS® clusters improves uniformity in the corners, but depending on the LED spacing and thickness of the backlight, it becomes more difficult to obtain good uniformity on the surface.

With LEDs, it is possible to create a solution based on best-cost and best-efficiency. Measurements are shown in Table 14.1. In the past, the LED cluster arrangement has been carried out solely in RGGB modules, but with increasing LED efficiencies, the option of RGB clusters has now become available as well.

14.4 Conclusions

For design activities, cost will be the first priority, and LED efficiency will improve year by year, as LEDs continue to achieve greater brightness. Reducing the number of components also leads to lower field-failure rates, reduced assembly costs and the easier driving of LED backlights. Even now, in the early stages of LED backlight technology for TV applications, the efficiency of mature CCFL technology can be easily achieved using LED backlight technology, and better image quality can be offered to customers.

Further improvements in efficiency due to an increase of LED brightness will no doubt be achieved in the near future. More benefits, such as improve-

Table 14.1 Measurement results of two different options for 32-inch LED backlighting compared to CCFL.

	LED (best cost)	LED (best efficiency)	CCFL (original)
cluster type	RGB	RGGB	–
number of LEDs	135	200	–
DC power consumption	138 W	105 W	110 W
luminance on LCD	511 nits	532 nits	450 nits
luminance on DBEF	6250 nits	6511 nits	5500 nits
uniformity	87%	88%	88%
CCT	9259 K	9214 K	12675 K
color coordinates	x = 0.284	x = 0.284	x = 0.275
	y = 0.296	y = 0.297	y = 0.267
	δx = 0.005	δx = 0.006	
	δy = 0.006	δy = 0.005	

ments in image quality due to special driving patterns for the LED backlight and sequential coloring, will accelerate the introduction of this new and exciting technology and bring better TV pictures to end consumers.

References

[1] Illek, S. et al. (2002) 'Buried micro-reflectors boost performance of AlGaInP LEDs', Compound Semiconductor, 8, pp. 39–42.
[2] Schmidt, W. et al. (2001) 'Efficient light-emitting diodes with radial outcoupling taper at 980 and 630 nm emission wavelength', Proc. SPIE LEDs: Research, Manufacturing and Applications V, 4278, pp. 109–118.
[3] Baur, J. et al. (2002) Phys. Stat. Sol. (a), 194, pp. 399–402.
[4] Kuhn, G. et al. (2005) 'A new light source for projection applications', SID '05 Digest, pp. 1702–1705.
[5] Ploetz, L. and Stich, A. (2005) 'LED Display Backlighting – Large Screen and TV Application using Golden DRAGON® ARGUS®', ApplicationNote.
[6] Schwedler, W. et al. (2005) 'LCD display backlighting solutions with high efficient surface emitting LED solutions', 20th Electronic Displays Conf., Session 3.5, Wiesbaden, Germany.
[7] Zeiler, M. et al. (2005) 'Highly efficient LED backlight solutions for large LCDs', 1st Crystal Valley Conference LCD Backlight '05, pp. 51–57.
[8] Kuhn, G. (2005) 'New LED lightsources for display technology', FPD International Forum.

15

Technological Trends of LED Backlight Units

Y. Kondo

NEC LCD Technologies

15.1 Introduction

Although CCFLs have been widely used as backlights for TFT-LCDs, other light sources are being investigated extensively. Among them, LEDs are experiencing significant performance improvements, enabling them to be used in relatively small-size displays. For portable devices, the use of white LEDs has become standard. Recently LC-TVs with LED backlights have been commercialized for medium-size devices.

15.2 Structure of LED Backlight Units

There are two types of backlight units as shown in Figure 15.1. One is a side-light type which diffuses the LED light by using a light-guide plate or a cavity. The other is a direct-light type in which LEDs are arranged at the back of the LC panel. LEDs are point light sources in contrast to the line light sources of CCFLs. In order to obtain a uniform distribution of light output, more careful designing of the light-guide plates, diffuser sheets, etc., is required for LED backlight units.

LCD Backlights   Edited by Shunsuke Kobayashi, Shigeo Mikoshiba and Sungkyoo Lim
© 2009 John Wiley & Sons, Ltd.

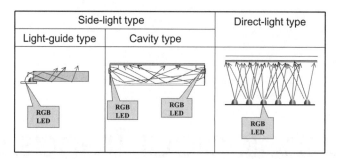

Figure 15.1 Various structures of LED backlights.

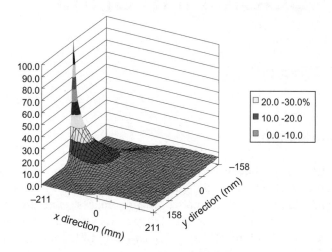

Figure 15.2 Distribution of light output from a light-guide plate type backlight unit when there is one LED light source.

15.3 Design of LED Backlight Units

15.3.1 Light-guide Type LED Backlight Unit

An LED light source package consists of three (R, G and B) LEDs. These packages are arranged linearly at a constant spacing. The LEDs are operated at a constant current. The distribution of light output from a light-guide plate is shown in Figure 15.2 when an LED light source is placed at $x = -211$ mm and $y = 0$. The light intensity decays to 10 % when the travel distance of the light becomes greater than 200 mm. The appearance of the light diffusion in the light-guide plate can be found in Figure 15.3. The light-guide type

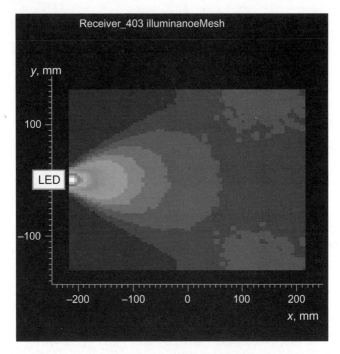

Figure 15.3   Diffusion of LED emission in a light-guide plate.

backlight unit generally utilizes two light-guide plates. The plate closer to the LEDs provides luminance and color uniformity while the second plate changes the point light source to a plane light source.

15.3.2 Cavity-type LED Backlight Unit

R, G and B rays are mixed in the cavity rather than in the light-guide plate. One of the features of the cavity-type backlight unit is its low weight. Also loss of light is less since, for the light-guide type, the light has to experience entering into and exiting from the plates twice; each of these processes has an associated optical loss. Also there are higher chances of reflection at the side walls for the light-guide plates.

The distribution of light output from the cavity-type backlight unit is shown in Figure 15.4. The decay of light intensity is more gradual than that of Figure 15.2. Figure 15.5 is an example of light-output distribution of the cavity type when LEDs are placed only at the bottom side. It can be seen that the luminance is uniform in the horizontal direction. By providing LEDs at both the top and bottom sides, vertical uniformity can be obtained (Figure 15.6).

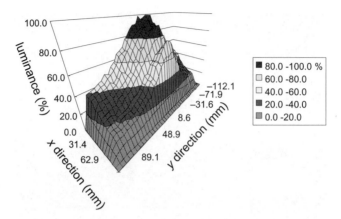

Figure 15.4 Distribution of light output from a cavity-type backlight unit when there is one LED light source.

LEDs

Figure 15.5 Luminance distribution of a cavity-type backlight; LEDs are placed at the bottom side.

15.3.3 Direct-light Type LED Backlight Unit

The direct-light type uses LEDs at the back of the LC panel. Loss of light is the smallest among the various types shown in Figure 15.1, but it is relatively difficult to achieve luminance and color uniformity. This is due to shorter paths of the light trajectories. For a light-guide plate, the light propagates a long distance along the plate surface and the optical path lengths become even longer by a factor of the refractive index of the plate. Also for the cavity

LEDs

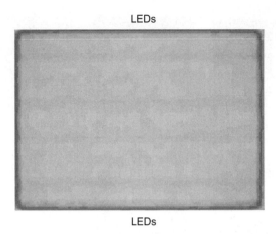

LEDs

Figure 15.6 Luminance distribution of a cavity-type backlight; LEDs are placed at both the top and bottom sides.

type, the light experiences multiple reflections, thus the optical paths are longer compared to those of the direct-light type.

For the side-light type, mixing of the colors is attained by the long optical path. This mechanism is not applicable to the direct-light type in which optical paths are approximately equal to the thickness of the backlight unit. The direct-light type therefore requires a different design such as inserting diffusive sheets between the light source and the LC panel. Figure 15.7 shows how the light diffuses with respect to the distance from the LEDs. At 10 mm above the LEDs, a circle of radius 22 mm is irradiated. The radius increases to 83 mm at a distance 50 mm from the LEDs. By overlapping the luminous regions, appropriate mixing of colors can be attained.

15.4 Requirements for Backlight Units

Requirements for the backlight units are determined by LCD modules and the products in which they are used. Among various items, the following requirements are of particular importance.

15.4.1 Luminance

The required backlight luminance is determined by the transmission of the LC panel. The luminous flux of each R, G and B LED is measured under a predetermined current level. The luminance fluctuates for each LED after

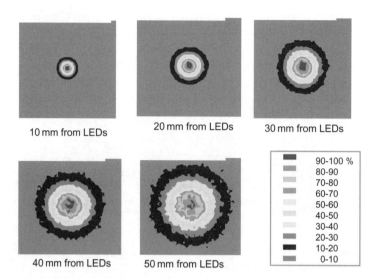

10 mm from LEDs   20 mm from LEDs   30 mm from LEDs

40 mm from LEDs   50 mm from LEDs

Figure 15.7 Diffusion of LED light; the vertical and horizontal sizes of the background squares are both 175 mm.

manufacture and hence suitable LEDs have to be selected before they are shipped to customers. Data of the average and lowest values should be accompanied when the LEDs are shipped.

15.4.2 Luminance Uniformity

The luminance uniformity of the backlight unit is defined by

luminance uniformity = (maximum luminance)/(minimum luminance),

where the maximum and minimum luminance are obtained by dividing the LCD screen into nine regions as shown in Figure 15.8, and measuring the luminance values at the centers of the regions (a)–(e). European standards divide the screen into 25 regions to obtain a more accurate value of uniformity.

15.4.3 Emission Spectra

Emission spectra also fluctuate for each LED after manufacture and hence suitable LEDs have to be selected. The measurement of the spectra is done under a predetermined current level and temperature. Data of the center, lowest and highest wavelengths should be accompanied with the shipment.

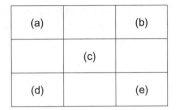

Figure 15.8 Measuring points for evaluating luminance uniformity.

15.4.4 Width of Spectrum

The width of the emission spectrum is also affected by the manufacturing process and the structure of LEDs. The HWHM (half-width at half-maximum) value should be specified.

15.4.5 Chromaticity and Color Gamut

The output colors are determined by the combination of the emission spectra of the LEDs and the transmission spectra of the color filters. The required CIE chromaticity coordinates of white, R, G and B for the LCD modules are specified by the customers. The color gamut can be expressed quantitatively by the area of a triangle made by RGB colors on the CIE1931 (x, y) chromaticity diagram. For more accurate expression, CIE1976 UCS (u', v') chromaticity coordinates should be employed. The CIE1976 allocates higher weights for R and B compared with CIE1931, more accurately matching the perception characteristics of the human eyes. For example, a typical CCFL has a CIE1931 area which is 73 % of the NTSC area, but a CIE1976 area is 83 % of NTSC. A typical LED backlight has a CIE1931 area which is 103 % of NTSC and a CIE 1976 area which is 115 % of NTSC.

15.4.6 Temperature Dependence

Emission spectra depend on the operating temperatures of LEDs. The dependence should be specified in terms of nm/°C.

15.4.7 Lifetime

Lifetime is defined by the time when the luminance is reduced to a half of the initial value. CCFLs have a life of 50 000 hours. The lifetime of LEDs depends on the operating current, temperature and ambient humidity, making it difficult to measure. Since the luminous efficiency of LEDs is lower than those of CCFLs, the operating temperatures of LEDs are generally higher. In addition, LED backlight units have a closed structure in order to admit LED light into the light-guide plate. The structure increases the temperature rise, leading to further shortening of the life. Also the lenses of LEDs are degraded by high temperature and humidity.

15.5 Technical Trends of LED Backlights

Backlights for cell phones once used inorganic EL. However, since their operating frequency is within the audible range, their luminous efficacy is low and their color representation is limited, inorganic EL has been replaced by white-emitting LEDs since the beginning of 2000. For medium-size LCDs, the backlight unit is not simply a light source; it also contributes to picture quality improvements including widening of the color gamut, reduction of motional blur and obtaining higher dark-room contrast. These improvements are attained mainly by the introduction of three-color (RGB) LED backlights. LCDs with LED backlights have been commercialized by Mitsubishi, NEC, Sony and other companies since 2002.

15.5.1 Wider Color Gamut

Figure 15.9 compares the emission spectra of a CCFL and an RGB-LED package. The LED has more intense spectra in the red region. Also it emits only pure R, G and B, providing high color purity. The color gamuts of a CCFL and an RGB-LED are compared in Figure 15.10. These have, respectively, 72 % and 103 % area with respect to the NTSC triangle.

15.5.2 Improvement of Emission Characteristics under Low Ambient Temperature

The temperature dependences of luminance and ignition voltage are illustrated in Figure 15.11 for a CCFL and an LED. For a CCFL, which utilizes mercury, the luminance is reduced to 20 % when the temperature is lowered from 25 °C to −20 °C. If the luminance of the backlight unit should be kept within ±20%, then the CCFL temperature should not fall below 10 °C. Also an ignition voltage one and a half times higher is required for the temperature variation from 25 °C to −20 °C. For LEDs the luminance and operating voltage are, much less dependent on the ambient temperature. For a range in temperature from −40 °C to +85 °C, luminance and operating voltage variations are kept within ±20% of the values at 25 °C.

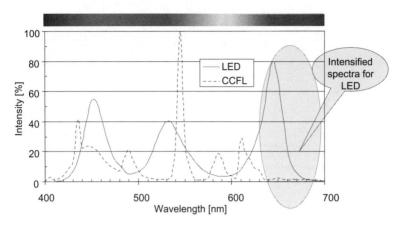

Figure 15.9 Emission spectra of a CCFL and an RGB-LED.

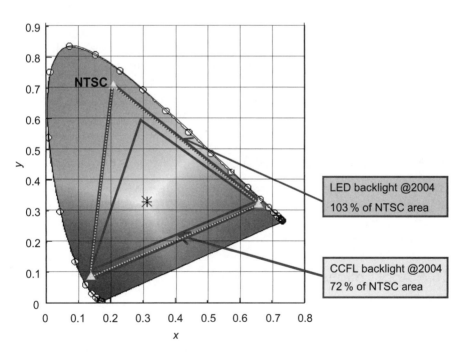

Figure 15.10 CIE chromaticity values for a CCFL and an RGB-LED.

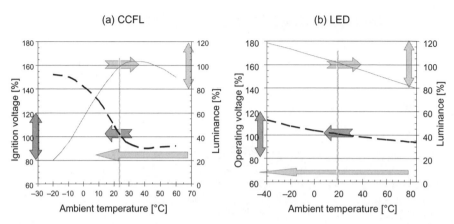

Figure 15.11 Temperature dependences of CCFL and LED characteristics.

15.5.3 Extension of Lifetime

The lifetime of an LED backlight unit is determined by the multiplication of LED emission degradation and optical component transmission degradation. For CCFLs, the life is reduced by too high and also too low temperatures. For LEDs, on the other hand, a low ambient temperature results in higher luminance at a constant current. This allows a reduction of the operating current which, in turn, reduces the LED temperature. As a result, the life is extended.

15.5.4 Picture Quality Improvement of Moving Images

Blinking backlights and local dimming techniques have been introduced to improve picture quality. The techniques require that the LEDs are switched on and off within 0.1 ms, which can easily be attained. Because of these advancements, a wider use of LED backlights is expected.

15.6 Applications of LED Backlights

15.6.1 For Moving Images

Moving pictures can best be shown on a CRT because of the short emission time of a pixel of the order of 1 μs. The LCD emission time is typically 17 ms (when the blinking backlight technique is not used), which is too long for obtaining the CRT-like images. The picture quality of LCDs can be improved by adopting the blinking backlight or a high frame frequency drive.

15.6.2 For Color Expression

The gamut of sRGB is approximately 72 % that of NTSC. To attain the Adobe RGB gamut, which is wider than that of sRGB, the use of LED backlights is necessary. In order to express all the colors found in nature, the image input device, image storage, image transmission, etc., also have to be improved.

In the field of medical electronics including film-less X-ray photography, it is essential to express images with stable white and blue colors. The RGB-LED backlight can adjust the white and blue colors arbitrarily for medical use. Luminance, color, temperature, etc., can be stabilized by a feedback loop of the output emission of LEDs.

15.7 Conclusions

The remaining issues for the development of LED backlight units are given below.

Cost reduction: Considering the additional costs of drivers and cooling fans, the cost of LED backlight units is appreciably higher than that of CCFL backlight units. A reduction in the cost of LEDs – as well as a reduction in the number of LEDs in a backlight unit – is necessary.

Lower power consumption: LEDs consume about twice as much power as CCFLs, indicating the need to improve the light emitting efficiency. Also losses arising from mixing the RGB light should be reduced.

Color non-uniformity: Color non-uniformity can be improved by reducing the distances between the R, G and B LEDs, although the heat dissipation becomes worse. Mixing of colors can alternatively be done by appropriately designing the backlight optics.

By improving LEDs and LED backlight unit structures, LEDs will be adopted not only for high-end LCD monitors, but for car electronics and home-use TVs.

16

White OLED Backlights

J. Jang

Kyung Hee University

16.1 Introduction

A white organic light emitting diode (referred to as a 'WOLED) is an OLED device which emits white light in an organic single- or multi-layer between cathode and anode electrodes. Research on WOLEDs started in the late 1980s, and a number of research projects are now pursuing possibilities of their application to OLED displays, LCD backlights and OLED lighting.

Many display corporations, such as Samsung Electronics, Eastman Kodak, Universal Display Corporation (UDC) and Sanyo, are now undertaking research and development to produce a full-color OLED display using WOLED technology. Toyota Industries Corporation in particular is now focusing its research and development, together with UDC, to apply WOLED technology to LCD backlights. Major lamp industries such as General Electric (GE) and Osram are interested in research and development to apply WOLEDs to general lighting. A WOLED can be made with small molecules or polymers and it produces fluorescent or phosphorescent emission. In addition, it can be made with a single-emission layer or multi-emission layers.

A number of structures are being proposed to create the WOLED: single-layer emission, multi-layer emission, color conversion, microcavity and tandem method. The important factors are opto-electrical properties related

LCD Backlights   Edited by Shunsuke Kobayashi, Shigeo Mikoshiba and Sungkyoo Lim
© 2009 John Wiley & Sons, Ltd.

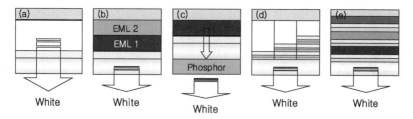

Figure 16.1 Methods for obtaining white emission from an OLED.

to color purity, color stability dependence upon current and voltage, luminous efficiency, lifetime and manufacturability. Figure 16.1 illustrates methods to obtain white emission from an OLED: (a) a single-layer emission which indicates that the white light can be achieved from a single layer, (b) a multi-layer luminescence with color combination by vertical integration, (c) a color conversion in which blue light is converted into white by using phosphor, (d) color conversion by a combination of blue and red phosphor and (e) tandem method whereby R, G and B OLEDs are integrated vertically.

16.2 White OLED with a Single-layer Emission

A polymer WOLED has low quantum efficiency because it has no functional layers such as hall injection layer (HIL), hall transport layer (HTL), electron transport layer (ETL), and color combination of R, G and B in a single layer. Therefore, it is hard to make such a WOLED with good color purity. The performance of polymer WOLEDs with a single-emission layer are collected in Table 16.1. In 1995, Yamagata University, Japan, fabricated a polymer WOLED with excellent color purity of mono-layer by utilizing PVK as a host, and a blue dopant 'TPB', green dopant 'coumarin 62', red dopant 'DCM1' and nile red. The maximum luminance was $4100 \, cd/m^2$. IMEC utilized PEB as a host, rubrene and PS as dopants, to make a single-layered WOLED emitting with a luminance value of $1500 \, cd/m^2$. A group in Technische University, Germany, achieved a quantum efficiency of 1.2 % in the case of WOLED doped with PPDB in m-LPPP.

A group at Kanazawa University in Japan developed a high-efficiency polymer WOLED in 2004. The WOLED was made by doping PVK with nile red, rubrene, TPB and Bu-PBD, and its power efficiency and color purity were $4.3 \, lm/W$ and $(0.33, 0.33)$, respectively. In the case of a polymer WOLED, it is difficult to achieve high power efficiency with just a single-layer. South China University increased the efficiency by spin-coating two polymer

Table 16.1 Performance of polymer WOLEDs with a single-emission layer.

Researcher	Materials	Luminance (cd/m²)	Efficiency	CIE	Reference
Yamagata Univ.	PVK, TPB, Coumarin 6, DCM 1, Nile Red	4100 (20 V)		0.33, 0.39	[1]
Kanazawa Institute	PVK, BBOT, Nile Red	5000	4.3 lm/W, 11.3 cd/A	0.33, 0.33	[2]
IMEC, Belgium	PEB:Rubrene: PS(16.6:0.07:83.3)	1500	0.9 cd/A, Q.E.:0.3%	0.30, 0.37	[3]
Technische Univ.	m-LPPP:PPDB		Q.E.:1.2%	0.33, 0.33	[4]
South China Univ. Tech	Blend 1(PVK + PFO-DHTBT)/Blend 2(PFO-poss P-PPV)	6300 (10 V)	4.4 cd/A	0.33, 0.32	[5]
NCTU, ROC	PFO + MEH/PFO	3000	1.6 cd/A	0.34, 0.34	[6]

layers. National Chiao Tung University (NCTU), Taiwan, successfully increased the device efficiency by two-layer coatings: a white luminescent layer and a blue emitting layer.

As in the polymer WOLED, a single-layered WOLED with small molecules uses dopants in a host material. A WOLED, doped with rubrene in a PAP-NPA host, showed brightness reaching 37 000 cd/m² with an efficiency of 2.51 lm/W at 5.57 cd/A, as listed in Table 16.2. A group at Jilin University in China has announced a WOLED with brightness of 29 000 cd/m² and an efficiency of 3.0 lm/W or 6.3 cd/A. ETRI in Korea has also announced a WOLED by doping DCJTB in SAlq which has a brightness of 20 400 cd/m², with an efficiency of 2.3 lm/W. A Shanghai University group fabricated a white OLED showing a brightness of 14 850 cd/m² with an efficiency of 2.88 lm/W, and 7822 cd/m² with 1.75 lm/W. In 1996, Bell Laboratory developed a single-layer luminescent white OLED which emits over a wide range of wavelengths, employing a substance called NAPOXA without doping. This was a single-layer luminescent device having a structure ITO/TAD/NAPOXA/Alq/Al, exhibiting white emission from NAPOXA. The brightness of this device was higher than 4500 cd/m² and its luminescent efficiency was 0.5 lm/W.

Table 16.2 Materials and performances of WOLEDs with a small-molecule, single-layered OLED.

Researcher	Materials	Luminance (cd/m²)	Efficiency	CIE	Reference
Academia Sinica	PAP-NP: Rubrene	37 000 (14 V)	2.51 lm/W, 5.57 cd/A	0.32, 0.34 (6~12 V)	[7]
Jilin Univ.	BePP2: Rubrene	29 000	3.0 lm/W, 6.3 cd/A	0.33, 0.33 (11 V)	[8]
ETRI	SAIq: DCJTB	20 400 (810 mA/m²)	2.3 lm/W, Q.E. = 2%	0.34, 0.39	[9]
Shanghai Univ.	JBEM(p): DCJTB	14 850	2.88 lm/W	0.32, 0.38	[10]
Shanghai Univ.	DPVBi: DCJTB	7822 (20 mA/cm²)	1.75 lm/W	0.25, 0.32	[11]
AT&T Bell Lab.	NAPOXA	4750	0.5 lm/W, Q.E.: 0.5%	0.31, 0.41	[12]
California Univ.	a-NPD, DPVBi, Rubrene, DCJTB, C545T	15 000 cd/m²	2.5 cd/A	0.31, 0.36	[13]
Tsing-Hua Univ.	D2NA: D2NBA: Rubrene	20 100	5.59 cd/A, Q.E. = 2.4%	0.31, 0.34	[14]

A group at the University of California developed a single-layer lumines-cent white OLED with an organics doped in a-NPD with DPVBi, rubrene, DCJTB and C545T. The group used Fused Organic Solid Solution which can precisely control dopants at high temperature and high pressure. The maximum luminance was 15 000 cd/m² and its efficiency was 2.5 cd/A. The color stability was excellent. Tsing Hua University, China, announced a white OLED with a power efficiency of 5.59 cd/A, and brightness of 20 100 cd/m² using D2NA, N2NBA dopants in rubrene.

16.3 White OLED with Multi-layer Emission

Table 16.3 explains various activities on WOLEDs with multi-layer emission. Although a polymer LED is constructed with multilayers only with diffi-culty, small molecules can easily be evaporated to make multilayers for

Table 16.3 Performance of WOLEDs using multi-layer emission.

Researcher	Materials	Luminance (cd/m^2)	Efficiency	CIE	Reference
Jilin Univ.	TBVB / Rubrene	4025	3.2 cd/A		[15]
Toyama Univ.	Rubrene / DPVBi	20 200	11.7 lm/W, Q.E. = 3%	0.26, 0.33	[16]
Ames Lab.	α-NPD:DCM2 / DPVBi	50 000	4.1 lm/W, Q.E. = 3.0%		[17]
Jilin Univ.	BePP2 / Rubrene	29 000	3.0 lm/W, 6.3 cd/A	0.33, 0.33 (11 V)	[18]
Inha Univ.	DPVBi / Alq3:Rubrene: DCM2	12 500 (7.4 V)	1.1 lm/W, Q.E. = 2.5%	0.34, 0.30	[19]
Academia Sinica	PAP-Ph / Alq / Alq3:DCM	24 700 (15 V)	1.93 lm/W	0.35, 0.34	[20]
ETRI	DPVBi / NPB:DADB / Alq	23 000	3.6 lm/W, Q.E. = 3.3%	0.34, 0.33	
Jilin Univ.	Bepp2 / BePP2:Rubrene / Alq3	20 000	1.11 lm/W	0.32, 0.33 (10 V)	[21]
I-Shou Univ.	Alq3 / Alq3:DCM2 / Anthracene:BczVBi	3750		0.33, 0.33	[22]
Princeton Univ.	NPD / NPD:DCM2 / BCP / Alq3		0.35 lm/W, Q.E. = 0.5%	0.33, 0.33	[23]
Jilin Univ.	Alq:DCJTB / BCP / Alq	6745	Q.E. = 1.36%	0.33, 0.33	[24]

higher emission. In 1999, Princeton University, USA, developed a new multi-layer luminescent WOLED. The a-NPD host doped with DCM2 emitted blue and red light in one layer and green light in an Alq ETL layer. Combining these emissions in the vertical device resulted in white light. The device, when BCP was inserted into an HBL layer, had a maximum luminance of 13 500 cd/m^2 and an efficiency of 0.35 lm/W, and demonstrated an excellent color stability against current variation.

Jilin University, China, created new devices by laminating TBVB and BePP2 with rubrene, which had brightness of 4025 cd/m^2 and 29 000 cd/m^2, and efficiencies of 3.2 cd/A and 6.3 cd/A, respectively. Toyama University, Japan, also announced its new high-efficiency WOLED, made by employing rubrene and DPVBi, with luminance of 20 200 cd/m^2 and an efficiency of 11.7 cd/A. The recent technology developed by Eastman Kodak was to combine yellow with blue light to obtain white light. Ruburene and Perylene were used for the yellow and blue luminescent layers, respectively.

Luminous efficiencies of OLEDs fabricated by employing phosphorescent materials is higher than that of fluorescent OLEDs, as shown in Table 16.4; accordingly, a lot of approaches have been used to achieve even higher efficiency with phosphorescent materials. Universal Display Corporation (UDC) and Toyota jointly announced a newly developed white OLED, which has an efficiency of 18.4 lm/W (39 cd/A), color coordinates of (0.39, 0.39), and CRI (Color Rendering Index) of 79. CRI is a measure of the quality of light, which indicates how the light shifts the color of objects. If there is no color shift, then the CRI of the light source is 100. The WOLED device was a multi-layer luminescent WOLED using R, G and B phosphorescent dopants. On the other hand, the phosphorescent WOLED reported by NHK and Tokyo University had a brightness of $18\,000\,cd/m^2$ and a luminous efficiency of 10 lm/W. NOVALED has announced a newly developed WOLED which

Table 16.4 Performances of WOLEDs with phosphorescent materials.

Researcher	Materials	Luminance (cd/m^2)	Efficiency	CIE	Reference
Jilin Univ.	Dmbpy-Re:CBP	2410	5.1 cd/A	0.30, 0.37	[25]
Yamagata Univ.	TPD/Tb(acac)3(Phen)/ Eu(DBM)3(Phen)			0.32, 0.33	[26]
UDC & Toyota	RD61/GD33/BD30	1000 (6.5 V)	38 cd/A, 18.4 lm/W	0.39, 0.39	[27]
Princeton Univ.	CBP:FPt2	31 000 (16.6 V)	9.2 cd/A, Q.E. = 4.0%	0.35, 0.43	[28]
California Univ.	PFO:Ir(HFP)3		4.3 cd/A	0.35, 0.38	[29]
NHK & Tokyo Univ.	CDBP:CF3ppy)2Ir(pic)/ BAIq/CDBP: Ir(btp)2(acac)	18 000	18 cd/A, 10 lm/W	0.35, 0.36	[30]
Princeton Univ.	Ir(ppz)3/CBP:PQIr/ UGH2:Ir(46dfppy)3		10 lm/W, Q.E. = 11%	0.42, 0.39	[31]
South China Univ.	PVk/ PFO:Ir(Bu- ppy)3:piq)2Lr(acaF)	10 200	9 cd/A, 5.5 lm/W	0.33, 0.33	[32]
Jilin Univ.	CBP:Ir(ppy)3: DCJTB/BCP	12 020 (19 V)	8.6 cd/A	0.33, 0.32	

has a CIE coordinate of (0.35, 0.37), a luminance of $1000\,cd/m^2$ and an efficiency of $16.3\,lm/W$. In collaboration with Philips, NOVALED recently announced their new WOLED developed by advanced PIN-OLED technology, which has a brightness of $1000\,cd/m^2$ at an efficiency of $25\,lm/W$. OSRAM also reported that the device using white-light emitting materials including phosphorescent emission material had an efficiency of $20\,lm/W$.

16.4 WOLED with Color Conversion

GE used a blue polymer OLED to make a high-efficiency WOLED with color-conversion, transforming blue into white light, as shown in Table 16.5. They used phosphors such as perylene dye with a quantum efficiency of 98 %, and Y(Gd)AG:Ce roughly 85 % respectively. The WOLED exhibited an efficiency of $16\,lm/W$ at a luminance of $1500\,cd/m^2$ $(19\,cd/A)$ and CIE coordinates of (0.36, 0.36). The device also showed $15\,lm/W$ at a luminance of $1000\,cd/m^2$ and a color temperature of $4400\,K$. OSRAM demonstrated a high-efficiency white OLED of $25\,lm/W$, using a high-efficiency blue polymer having an efficiency of $12\,lm/W$ together with a color-conversion layer.

16.5 Stacked WOLED Devices

Table 16.6 lists the performances of stacked WOLED devices. Princeton University, USA, demonstrated a newly developed OLED (stacked OLED or SOLED) using a stacked structure, with which independent OLEDs such as R, G and B are built, and then mixed into white. In addition, the stacked OLED has benefits from its process that the brightness of R, G and B can easily be controlled independently to obtain white emission. International Manufacturing and Engineering Services (IMES) reported its findings on an

Table 16.5 Performances of WOLEDs using color conversion.

Researcher	Materials	Luminance (cd/m^2)	Efficiency	CIE	Reference
G.E.	PF based blue LEP	1080	$3.76\,lm/W$, $6.57\,cd/A$		[33]
G.E.	PF based blue LEP	1000	$19\,cd/A$, $15\,lm/W$	0.36, 0.36	[34]
NNL of Italy	TPD/STO/ T5oCx			0.31, 0.34	[35]

Table 16.6 Performance of stacked WOLEDs devices.

Researcher	Materials	Luminance (cd/m²)	Efficiency	CIE	Reference
Princeton Univ.	Ir(ppz)3/mCP:Ir(flz)3/ CBP:Ir(ppy)3:PQIr		77 cd/A, 23 lm/W, Q.E. = 35%	0.35, 0.44	[36]
Princeton Univ.	Alq3/Alq2′OPH/ Alq3:DCM2	100 (15 V)		0.32, 0.34	[37]
IMES	Spiro-DPVBi(blue)/ CGL/Alq:DCJTB(red)			0.32, 0.34	[38]

OLED device employing a multi-photon emission (MPE) method. In this method, R, G and B devices are stacked consecutively, and then a charge generation layer (CGL) is inserted between each layer to increase luminous efficiency. The panel was reported to have color coordinates of (0.32, 0.34).

16.6 Applications of WOLEDs

16.6.1 General Lighting

The value of the worldwide market for general lighting is estimated to be US$80 billion. Multinational corporations including Philips of the Netherlands and GE of USA have a significant portion of the world's lighting market. With a continuous decrease in renewable energy resources and increases in oil prices, OLED and LED advanced lighting devices are expected as technologies that satisfy both an environment-friendly industrial structure and high power efficiency. To develop these light sources, the US, European and Japanese governments are investing substantial amounts of money into WOLED research.

Edison's first light bulb had an efficiency of about 15 lm/W, followed by the invention of a more efficient fluorescent lamp with an efficiency of 50 lm/W. Despite an improvement in its efficiency, the color rendering of the fluorescent lamp has not as yet improved significantly. The CRI of the incandescent light bulb is roughly 100, compared to the fluorescent lamp of 75. It is expected that an OLED will have a CRI higher than 85 by 2010, so that it can be used for general lighting applications. It therefore is suggested that LEDs and OLEDs could have great potential as spot and areal light sources, respectively. It is expected that lighting applications using WOLEDs will be a hot topic because of its potential.

Since a WOLED is thin and has high-brightness, it can readily be used in a backlight unit. Also its emission of natural light makes it comfortable to use. The manufacturing process of a WOLED for a backlight unit is simple, so it is relatively easy to adopt them for large-area displays. Efficiencies of lamps used for general lighting are increasing continuously. Metal halide lamps using high intensity discharges are still one of the most efficient lamps with an efficiency of 100 lm/W. The efficiency of WOLEDs has improved significantly to the value of 25 lm/W. WOLEDs have a high possibility of being used in light applications since the incandescent lamp has an efficiency level of only 15 lm/W.

16.6.2 LCD Backlights

White OLEDs can be adopted in LCD backlight units because of their thinness and high brightness. Their manufacturing process is simple and thus can be widely used without any sophisticated pattern-forming processes. In addition, the white OLED, as an area light source, has excellent brightness uniformity, compared to the line light source of a CCFL or the spot light source of an LED. Moreover, the number of components can be reduced since a light-guide plate, prism sheet, etc. are not necessary.

OLEDs are required to have a luminance of $2000 \, cd/m^2$ to be adopted for backlights of small LCDs for mobile applications, and $10\,000 \, cd/m^2$ for backlights of large LCDs for TVs. The existing CCFL can hardly be adopted for small size LCDs like cell phones due to the limited space and thickness. In addition, LC-TVs require a large number of CCFLs and inverters. Consequently, the cost of LC-TVs increases. Also color reproducibility of LC-TVs is limited because of the color reproducibility of CCFLs. Tens or hundreds of LEDs are needed for constructing a backlight of an LC-TV because there is a limit to the amount of light flux generated from each LED. If white OLEDs are adopted for backlights, they can create an area light source with high-brightness even with low voltages.

16.7 Research and Development Status

Since 2000 the US Department of Energy (DOE) has been conducting a project related to SSL (solid-state lighting) aimed at developing a light source with high power efficiency. The entire project, with more than US$63 million in funding, focuses on developing low-cost LEDs and OLEDs using various methods. The project includes OLED-related projects that are estimated to cost around US$29.8 million. Research and development programs are now under way, conducted by many leading companies including UDC and GE. The USA is continually investing a substantial amount of money into the WOLED industry. If successfully developed, the low-cost and highly energy-efficient OLED will have a great potential in terms of its environmental preservation and advanced lighting applications.

Funded by DOE, UDC is performing a leading-edge research and development project on white-light materials with excellent power efficiency and device configuration as alternatives to incandescent bulbs and fluorescent lamps, based on its phosphorescent OLED material technology. UDC is conducting a joint-development of WOLED (PHOLED) with Toyota to develop white lighting using RGB with low cost and high power efficiency. The 6 × 6 inch prototype of the WOLED was shown at SID '04 after improving phosphorescent emissive materials, driver design and fabrication methods. The panel has a luminance of $1000\,cd/m^2$ ($18\,lm/W$) and color coordinates of (0.38, 0.38) at a driving voltage of 6.3 V.

Furthermore, UDC demonstrated a new 6 × 6 inch WOLED lighting panel having a power efficiency of $30\,lm/W$, using phosphorescent OLED technology at the Society of Optical Engineering Symposium and Exhibition in 2005. The OLED panel showed the highest power efficiency so far, with a similar level of color temperature (4000 K) fluorescent lamp. In addition, UDC developed a new phosphorescent OLED for low voltage operation, and n/p-type electrode or hole-moving material with high-conductivity. As a result, a PIN-type white PHOLED light source has been obtained by employing this material, which had color coordinates of (0.39, 0.39) and a luminance of $800\,cd/m^2$ ($19.7\,lm/W$ at 6.3 V). The light efficiency needs to be improved so as to break into the lighting business market, since only 25 % of the photons are extracted out of the device. As part of an effort to increase the light efficiency, UDC developed a new light extraction technology, called micro lens silicon mold with PDMS (poly-di-methyl-siloxane) and increased the out-coupling efficiency through adjustment to its shape and distance.

The OLLA project is an integrated research and development project funded by the European Commission under the IST (Information Society Technology) initiative, which started in 2004. More than 20 European companies are jointly developing high brightness white OLED light tiles for use, with a long lifetime and high efficiency. Universities and research corporations in Europe, including the companies Aixtron, Covion, Novaled, Osram, Philips, Sensient and Siemens, have joined the project. The first project goal in 2006 was to develop a white OLED having a lifetime of 2000 hours with an efficiency of $10\,lm/W$ at $1000\,cd/m^2$, and by 2008 to develop a high-brightness white OLED aiming at a lifetime of 10 000 hours, which is 10 times longer than the incandescent bulbs, with an efficiency of 50 lm/W.

The development of OLEDs for general lighting applications in Japan is being carried out by collaboration between research centers and several enterprises. The main objective is to develop a new lamp having both luminescent efficiency and long lifetime with almost the same levels of incandescent light bulbs by 2007, and fluorescent lamps by 2010. IMES company demonstrated two types of white OLED products, one of which having 400 cd/m² was a WOLED (28 × 28 cm) using a lime-green and orange color. The other (6 cm × 8 cm) used four luminescent layers emitting blue, green, yellow and red, and had luminance of 3000 cd/m² with color reproducibility of 94 %, which is a similar level to that of CCFLs. Tohoku Device Co., Ltd is developing a WOLED which uses phosphor materials of low-polymer. Its luminance is expected to be 1000 cd/m², and lifetime to be 10 000 hours.

Figure 16.2 shows the representative data for the luminescence and power efficiency of the WOLEDs developed so far. Power efficiency is increasing continuously and has reached 30 lm/W. This was achieved by adopting phosphorescent material by UDC,[42] GE increased the power efficiency by using down conversion. UDC and Princeton University focused on the use of PHOLEDs. NOVALED improved the power efficiency with the use of a *p-i-n* structure. By improving the device structure and materials, the efficiency and brightness of WOLEDs are expected to increase continuously. The lifetime will also be improved.

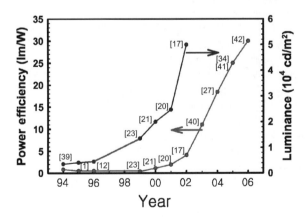

Figure 16.2 Power efficiency and luminescence of WOLEDs plotted by year; the numbers in the diagram indicate references.

References

[1] Kido, J. *et al.* (1995) *Appl. Phys. Lett.*, **67**, p. 2281.

[2] Mikami, A. *et al.* (2004) *SID '04*, p. 146.

[3] Yang, J. P. *et al.* (2000) *Chem. Phys. Lett.*, **325**, p. 251.

[4] Tasch, S. *et al.* (1997) *Appl. Phys. Lett.*, **71**, p. 2883.

[5] Xu, Y. *et al.* (2005) *Appl. Phys. Lett.*, **86**, p. 163502.

[6] Ho, G. K. *et al.* (2004) *Appl. Phys. Lett.*, **85**, p. 20.

[7] Chuen, C. H. and Tao, Y. T. (2002) *Appl. Phys. Lett.*, **81**, p. 4499.

[8] Duan, Y. *et al.* (2004) *Jpn. J. Appl. Phys.*, **43**, p. 7501.

[9] Ko, Y. W. *et al.* (2003) *Thin Solid Film*, **426**, p. 246.

[10] Jiang, X. Y. *et al.* (2002) *Synth. Metal.*, **129**, p. 9.

[11] Zheng, X. Y. *et al.* (2003) *Display*, **24**, p. 121.

[12] Jordan, R. H. *et al.* (1996) *Appl. Phys. Lett.*, **368**, p. 1192.

[13] Shao, Y. and Yang, Y. (2005) *Appl. Phys. Lett.*, **86**, p. 073510.

[14] Wang, L. *et al.* (2004) *Jpn. J. Appl. Phys.*, **43**, p. 560.

[15] Cheng, G. *et al.* (2004) *Appl. Phys. Lett.*, **84**, p. 4457.

[16] Tsuji, T. *et al.* (2005) *Current Applied Physics*, **5**, p. 1.

[17] Cheon, K. O. and Shinar, J. (2002) *Appl. Phys. Lett.*, **81**, p. 1738.

[18] Duan, Y. *et al.* (2004) *Jpn. J. Appl. Phys.*, **43**, p. 7501.

[19] Lee, M. J. *et al.* (2003) *SID '03*, pp. 525–528.

[20] Ko, C. W. and Tao, Y. T. (2001) *Appl. Phys. Lett.*, **79**, p. 4234.

[21] Liu, S. *et al.* (2000) *Thin Solid Films*, **363**, p. 294.

[22] Chen, S. F. and Wang, C. W. (2003) *SID '03*, pp. 521–524.

[23] Deshpande, R. S. *et al.* (1999) *Appl. Phys. Lett.*, **75**, p. 888.

[24] Cheng, G. *et al.* (2004) *Thin Solid Films*, **467**, p. 231.

[25] Li, F. *et al.* (2003) *Appl. Phys. Lett.*, **83**, p. 4716.

[26] Kido, J. *et al.* (1996) *Jpn. J. Appl. Phys.*, **35**, p. 394.

[27] Tung, Y. J. *et al.* (2004) *SID '04*, pp. 48–51.

[28] Deshpande, R. S. *et al.* (2002) *Adv. Mater.*, **14**, p. 15.

[29] Gong, X. *et al.* (2004) *Adv. Mater.*, **16**, p. 615.

[30] Tokito, S. *et al.* (2005) *Current Applied Physics*, **5**, p. 331.

[31] Kanno, H. *et al.* (2005) *Appl. Phys. Lett.*, **86**, p. 263502.

[32] Xu, Y. *et al.* (2005) *Appl. Phys. Lett.*, **87**, p. 193502.

[33] Duggal, A. R. *et al.* (2002) *Appl. Phys. Lett.*, **80**, p. 3470.

[34] Duggal, A. R. *et al.* (2005) *SID '05*, pp. 28–31.

[35] Mazzeo, M. *et al.* (2003) *Synth. Metal.*, **139**, p. 675.

[36] Kanno, H. *et al.* (2006) *Adv. Mater.*, **18**, p. 339.

[37] Burrows, P. E. *et al.* (1998) *Appl. Phys. Lett.*, **73**, p. 435.

[38] Matsumoto, T. *et al.* (2003) *SID '03*, pp. 979–982.

[39] Kido, J. *et al.* (1994) *Appl. Phys. Lett.*, **64**, p. 815.

[40] Brian, W. D. *et al.* (2003) *SID '03*, pp. 967–970.

[41] Werner, A. *et al.* (2006) *SID '06*, pp. 1099–1102.

[42] www.ledsmagazine.com 2 Aug., 2005.

17

Inorganic EL Backlights

S. Okamoto

NHK Science & Technical Research Laboratories

17.1 Introduction

An inorganic electro-luminescence (EL) device is an all-solid type that emits light by applying a high voltage. The device consists of fluorescent powder or a fluorescent thin film. During the period 1950–1970, active research and development was conducted aimed at studying the operating mechanism, improvements of luminance and lifetime by using the fluorescent powder type, but without much success. In the 1970s, the interest of researchers shifted to thin-film EL devices that ensured a lifetime of several tens of thousands of hours; more recently full color thin-film EL devices have been developed for information displays.

In the meantime, powder type EL devices were developed and used for backlights, watches and electronic ornaments by using their advantage of flexibility. The total sales volume, however, was not large. Recently demand for LCDs, particularly for TV use, has been growing fast demanding high quality backlights. As a solution to this, an inorganic EL device was developed which was capable of delivering $350\,000\,cd/m^2$ luminance with a lifetime of 25 000 hours.[1] Various types, structures, characteristics and light-emitting materials of inorganic EL devices will be introduced in this chapter.

LCD Backlights Edited by Shunsuke Kobayashi, Shigeo Mikoshiba and Sungkyoo Lim

17.2 Classification of Inorganic EL Devices

Inorganic EL devices can be classified as dispersion (powder) types or thin-film types, as shown in Figure 17.1. Further, each of these is divided into an AC type and a DC type. Powders for the EL devices are synthesized under a thermal equilibrium condition, while thin films are formed on a glass substrate in a non-equilibrium condition at an appropriate temperature. A dispersion type driven by an AC voltage is called an EL sheet. The device does not contain toxic elements, such as lead or mercury, and can be used as a backlight for a flexible LCD. On the other hand, the thin-film type EL devices driven by an AC voltage have the feature of long lifetimes, and they have frequently been installed in manufacturing apparatus as yellow color flat-panel displays.

DC EL devices of both the dispersion and thin-film type were once intensively developed as displays for automobile dashboards.[2] Due to their lack of reliability and short lifetime, however, they were not put into practical use. For this reason, research and development on EL devices almost ceased.

Emission from DC and AC inorganic EL devices is generated by accelerated carriers under an intense electric field. An LED is not included in Figure 17.1 since its light emission originates from the recombination of injected carriers.

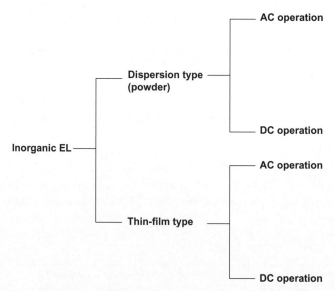

Figure 17.1 Various types of inorganic EL devices.

17.3 Device Structures and Characteristics

17.3.1 Dispersion-type ELs

17.3.1.1 Device Structure

The basic structure of an AC dispersion type inorganic EL device is shown in Figure 17.2. A light emitting layer with a thickness of 10–50 μm is formed by dispersing ZnS phosphor, which is doped with Cu and other elements, into a binder. One of the sides of the layer is then covered with a dielectric layer of BaTiO$_3$. The light-emitting layer and dielectric layer are sandwiched between two electrodes. The electrode attached to the light-emitting layer is optically transparent. The structure is packaged between films such as PET, so that it can be used as a flexible light source. ZnS phosphor particles are coated with a dielectric film such as silica in order to protect them from water damage. The thickness of the EL sheet is 0.2–0.7 mm, and the minimum bending radius is about 5 mm. There is no deterioration of performance as a result of bending the display multiple times.

17.3.1.2 Characteristics

Luminance L and applied voltage V are related by

$$L = L_0 \exp\left[-\left(\frac{V_0}{V}\right)^{1/2}\right], \tag{17.1}$$

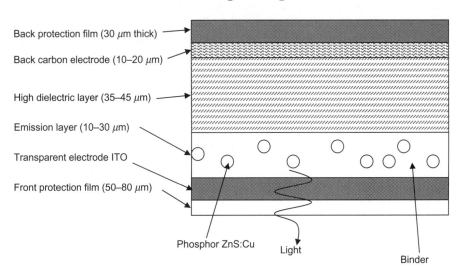

Back protection film (30 μm thick)

Back carbon electrode (10–20 μm)

High dielectric layer (35–45 μm)

Emission layer (10–30 μm)

Transparent electrode ITO

Front protection film (50–80 μm)

Phosphor ZnS:Cu   Light

Binder

Figure 17.2 Schematic cross section of an AC powder EL device.

where L_0 and V_0 are constants. An $L-V$ curve of a blue-green EL is shown in Figure 17.3(a). If the graph is re-plotted, as in Figure 17.3(b), an almost straight line is obtained. The luminance increases with increasing frequency up to 10 kHz. Typical performance is $100 \, cd/m^2$ at $V = 200-300 \, V$, $400 \, Hz$ and $0.1-0.3 \, mA/cm^2$. The EL sheet has a capacitance of several hundreds of pF/m^2. Luminous efficacy is typically less than $10 \, lm/W$. There was a recent report that $15-20 \, lm/W$ was obtained at $300 \, cd/m^2$. The half luminance decay lifetime is 1000–5000 hours for devices having an initial luminance of $100-3000 \, cd/m^2$.

17.3.1.3 Mechanism of Light Emission

Fischer's model[3] for the mechanism of the light emission of inorganic dispersion-type EL devices is considered to be very reasonable, and therefore no additional research has been done. Figure 17.4 illustrates Fischer's model, where it is assumed that there are line defects within ZnS particles and Cu_xS is formed in these regions by precipitation of Cu ions. Cu_xS has a p-type carrier conduction and it forms a heterojunction with n-ZnS. When an electric field of the order of $10^8 \, V/m$ is applied across the junction, electrons are injected into the ZnS layer by a tunneling effect. The Cu ions play the role of acceptors in the ZnS, and so a recombination with trapped holes occurs, and light is emitted. Since electrons and holes are separated under a high applied electric field, a sufficient amount of space and time are necessary for the recombination to take place.

Light emission occurs twice in every half cycle of the AC voltage and there is a delay between the applied voltage and the emission. For the material ZnS:Mn,Cu, it is generally believed that emission occurs from direct collisions of injected electrons with light-emitting center, an Mn^{2+} ion, that has an inner shell electron transition.

17.3.1.4 Light-emitting Materials

ZnS is always used as a host material. Light emission ranging from blue-green to green is obtained by doping Cu ions as acceptors and also I, Cl, Al, etc., as donors.[4] Especially when Cu, Cl and Mn are doped simultaneously, a yellow-orange emission from Mn^{2+} ions is obtained. The size of the ZnS particles is about $10 \, \mu m$; this is larger than those used in, for example, CRTs. The spectrum of a ZnS:Cu EL device emitting blue-green color is shown in Figure 17.5. By mixing red-emitting dyes, white light is obtained whose spectrum is shown in Figure 17.6. The color coordinates of the white color are (0.29, 0.31).

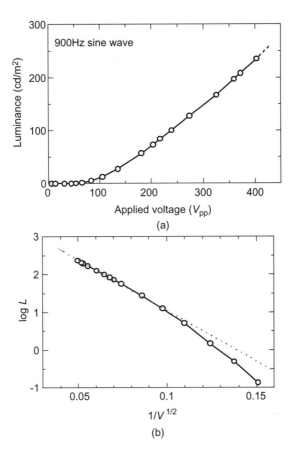

Figure 17.3 (a) Luminance versus applied voltage in a dispersion-type EL device. (b) *L–V* characteristic plot.

17.3.1.5 DC EL Devices

DC EL devices have a structure shown in Figure 17.7. A thin film, 50 μm thick, is formed on a transparent ITO electrode. The thin film is formed by mixing phosphor powders and an organic binder. The phosphor powder used in this device is ZnS:Cu, Mn which is formed by immersing ZnS powder in a copper sulfate solution. The immersion process is called the Cu coating of ZnS. The density of the phosphor is high compared with that of an AC type. Finally, an electrode is form on the Cu-coated ZnS layer. When a DC voltage of several tens of volts is applied, at first no light emission is observed. If the DC voltage is gradually increased, then light emission is

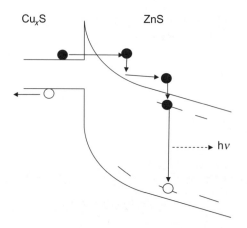

Figure 17.4 Mechanism of excitation and photon emission in a dispersion-type EL device.

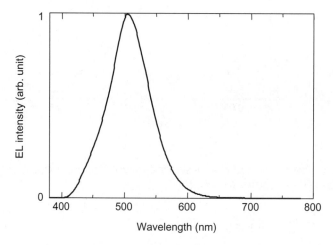

Figure 17.5 EL spectrum of a dispersion-type EL device using ZnS:Cu.

obtained.[2,5] Through this process, called the forming process, the coated layer of Cu_xS on the ZnS particles migrates from the positively biased ITO electrode to develop a highly resistive layer of ZnS.

A high DC electric field is applied across the forming layer and the Cu-coated ZnS layer. This field produces electron injection from the highly electrically conductive Cu_xS into the ZnS layer, and the electrons are acceler-

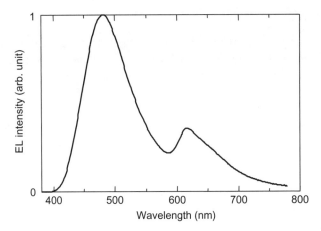

Figure 17.6   EL spectrum of a white light dispersion-type EL device.

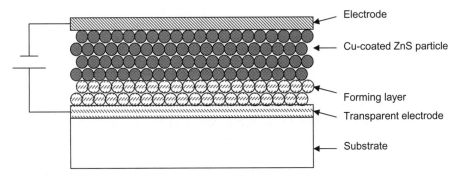

Figure 17.7   Schematic cross section of a DC powder EL device.

ated and collide with light emitting Mn ions. The maximum luminance achieved is about $1000\,cd/m^2$ and the $L–V$ characteristics almost obey Equation (17.1). DC EL devices have a half luminance lifetime of about 1000 hours. Besides ZnS:Mn,Cu, there are several other types using rare earth ion activators such as Re, (e and Eu) to create ZnS:Re,Li(Na), SrS:Ce,Cl, CaS:Ce,Cl or CaS:Eu,Cl that enable DC EL devices to emit colors from blue to red.

17.3.2 Thin-film Type EL

Compared with the DC driven dispersion type, thin-film EL devices driven by a DC voltage are regarded as EL devices that have a higher density of the emitting layer. The stacking of a resistance layer, composed of a Ni doped SiO_2 cermet, is an effective way to overcome the occurrence of electrical breakdown in thin-film EL devices. A high-luminance inorganic EL device, which is fabricated by vacuum evaporation, has a lifetime of 25 000 hours.

An AC driven thin-film EL device has a structure in which a light-emitting film insulating layer(s) has attached to one or both sides. This structure suppresses electrical breakdown and yields a long lifetime. The mechanism of the light emission of thin-film EL devices is attributed to the impact of highly accelerated electrons on the light-emitting centers. These types of EL panel are not used as backlights but they have been adopted for large area TVs with a high resolution X–Y matrix.[6–8]

17.4 High-luminance Inorganic EL Devices

T. Chatani and Co., Ltd, developed a high-luminance EL device emitting blue light at 350 000 cd/m^2 with a lifetime of 25 000 hours through a collaboration with Kuraray Co., Ltd.[1] The device had a threshold voltage of 3 V DC and a luminance of 600 000 cd/m^2 with 5.5 V and 2 000 000 cd/m^2 with 8 V. The specifications of the prototype were 2 mm square with an electrical current of several tens of mA, and luminous efficiency of 10–15 lm/W. The prototype was fabricated by vacuum deposition. Figure 17.8 shows the Chatani panel which was capable of emitting white light by combining blue with a YAG:Ce phosphor.

Various research has been conducted aiming at developing white-emitting EL devices for LCD backlights. The characteristics of a high-luminance EL device are shown in Table 17.1 comparing them with those of conventional devices. Compared with LEDs, ELs have advantages such as low production costs, since there is no need for epitaxial growth. For powder ELs, a simple coating of the relevant materials on a film is enough to fabricate flexible devices. An area light source does not require peripheral films or plates, unlike CCFLs and LEDs. This simplifies the structure of conventional LCDs and brings the total cost down. It is necessary to carry out further research and development on improving color capability and luminous efficiency by introducing new binders for the phosphor deposition. Investigation of methods for driving a large area DC EL device is also necessary.

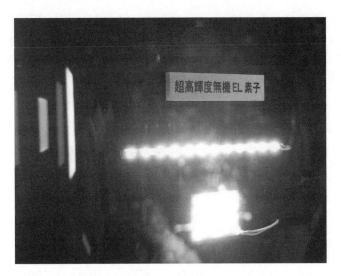

Figure 17.8 Photograph of a white light emission from an EL device using a super high-luminance blue inorganic EL device.

Table 17.1 Characteristics of a super high-luminance inorganic EL.

	Luminance (cd/m²)	Lifetime(hours)
Super high-luminance inorganic blue EL material	350 000	No degradation for 25 000 hours
Conventional inorganic blue EL material	100	Half value decay time 30 000 hours
Conventional organic EL material	1000	Half value decay time 10 000 hours

17.5 Practical Examples of Backlight Use

The first notebook PCs – Dynabook J-3100 SS001 – were put on the market by Toshiba in 1989. The notebooks were installed with backlights adopting AC driven dispersion-type inorganic EL panels. The specifications of these notebook PCs were 640 × 400 pixels with blue-white monochromatic emission. Although these notebook PCs were ground breaking, the EL backlights were replaced by CCFL tubes shortly afterwards. However, EL backlights are still used for wristwatches, Model Baby G and G-shock by CASIO, and a stick controller of the SONY MD/CD Walkman. They are also used for backlights of mobile phones and uniformly lighting the keypads, where an advantage of the thinness of the thin-film EL device appeals to the public.

For driving EL devices, CMOS transistors are commonly used due to their advantage of tolerance to a high voltage. The specifications are: voltage source 1.6–5.5 V, EL operating voltage 200 V_{pp}, frequency 30–1500 Hz and current 100 mA. It is possible to drive an EL sheet with an area of several tens of square centimeters with these specifications.

References

[1] Chatani and Kuraray release, October 12, 2005, http://www.chatani.co.jp/jptop/press/press.html.
[2] Vecht, A. (1982) 'Development in electroluminescence panels', *J. Crystal Growth*, **59**, pp. 81–97.
[3] Fischer, A. G. (1966) *Luminescence of Inorganic Solids*, Academic Press, p. 541.
[4] Shionoya, S. and Yen, W. M. (eds) (1999) *Phosphor Handbook*, Place: CRC Press, pp. 581 and 601.
[5] Dean, P. J. (1981) 'Compositions and contrasts between light emitting diodes and high field luminescence devices', *J. Luminescence*, **23**, pp. 17–53.
[6] Wu, X. *et al.* (2005) *Proc. 12th IDW/AD' 05*, pp. 1109–1112.
[7] Tanaka, K. and Okamoto, S. (2005) *Proc. 12th IDW/AD '05*, pp. 1625–1628.
[8] Miura, N. *et al.* (2005) *Oyobuturi*, **74**, p. 617 (in Japanese).

18

Field Emission Backlights

M. Ushirozawa

NHK Science & Technical Research Laboratories

18.1 Introduction

A field emission backlight is a backlight that uses a field emission lamp (FEL). The lamp generates electrons by field emission, accelerates them in a vacuum using an electrical field and emits light with cathodoluminescent phosphor. The lamp, with a vacuum space between an electron emitter (the cathode) and a phosphor screen (the anode) as shown in Figure 18.1, can be driven by simply applying a DC voltage. A feature of FELs is that the structure is extremely simple, as long as the diagonal dimension is not too large so that the effect of atmospheric pressure on the lamp container need not be considered. In this chapter we will give an overview of the primary structural elements of FELs, discuss recent research trends and follow this with a summary of issues relating to the successful production of field emission backlights.

18.2 Field Electron Emitter

When a high electrical potential of several MV per mm is applied to a metal or semiconductor surface, electrons on the surface escape the potential barrier due to the tunneling effect and are emitted from the surface. This effect is called the field emission effect and it has been well understood for

LCD Backlights   Edited by Shunsuke Kobayashi, Shigeo Mikoshiba and Sungkyoo Lim
© 2009 John Wiley & Sons, Ltd.

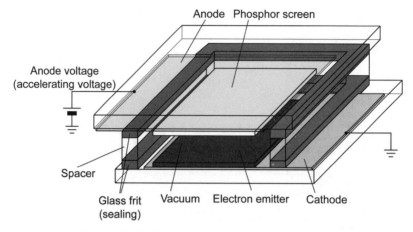

Figure 18.1   Diode-type field emission lamp.

some time.[1] The power required for field emission is extremely small compared with that used for hot-cathode emission in cathode-ray tubes (CRTs). Only the power efficiency of the phosphor screen need be considered when calculating the power consumption of an FEL.

In practice, it is quite difficult to obtain field emission from a smooth solid surface by simply applying a high potential. However, if there are irregularities on the surface, localized areas of high potential can be generated by the so-called shape effect.[2] Among the fibrous materials consisting of carbon atoms, such as carbon nanotubes (CNTs), there are fibers of diameter from several nm to several tens of nm and lengths of several μm. These materials can yield field-enhancement factors (the local field strength relative to the average field strength) of the order of hundreds or even thousands. This allows electron emission to be induced even with an average potential of only several KV/mm.[3] There are various methods for producing CNTs, including arc discharge, laser ablation and chemical vapor deposition (CVD).[4] The CNTs can be formed on a substrate using methods such as screen printing.[5–7] CNTs can also be formed directly on a substrate by CVD.[8] Regardless of the method used, it is relatively easy to form CNTs over a large area.

The simple FEL structure shown in Figure 18.1 has two electrodes: the anode, which is the phosphor screen, and the cathode which is the electron emitter. The anode provides the electron accelerating potential. As an example, to achieve an average fieldstrength of 5KV/mm with an anode voltage of 5kV, the distance between the anode and cathode should be 1mm.

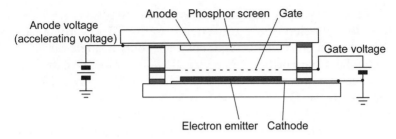

Figure 18.2 Triode-type field emission lamp.

18.3 Lamp Container and Vacuum Seal

The field emission lamp container must withstand atmospheric pressure. There have been reports of lamp containers for field emission displays (FED) up to 40-inch diagonal that can withstand this pressure without spacers.[9] Generally, however, 5-inch diagonal lamps are the maximum limit,[10] if spacers are not used. Research into the fabrication of spacers is advancing.[11,12] One approach is to make spacers using the thick-film screen-printing technique.[7]

In order to produce a lamp container consisting of several parts, it must go through a vacuum-sealing process. There is a long-established method of using glass frit which is inexpensive and has only a small amount of gas emission over an extended period. The method, however, requires a high-temperature process in the atmosphere, which degrades the properties of the carbon compounds. The frit-seal method in an argon[5] or nitrogen environment has been proposed.[13]

A triode type has also been proposed because it provides design freedom and can be driven easily. As shown in Figure 18.2, the structure includes an extraction/gate electrode placed near the cathode between the anode and cathode. This electrode induces the desired electrical field at the cathode to extract electrons. Unlike the diode structure, the voltage applied to the anode, namely the acceleration voltage, can be chosen arbitrarily. Because the gate voltage is much lower than the anode voltage, it is easy to switch the backlight on and off and to adjust its brightness.

18.4 Cathodoluminescent Phosphor

Development of cathodoluminescent phosphors has not always been for FELs for backlight units. The most well-known cathodoluminescent phosphor is P-22, which is used in CRTs and is a sulfide-type phosphor.[14] The wavelength of the emission nearly corresponds to the NTSC chromaticity point and the emission efficiency is high. For CRTs, only a single pixel is excited at a time. The instantaneous luminance, therefore, must be extremely high. For instance, if a phosphor screen is excited at 30 KV, a high electron-beam current density of 130 A/m^2 is required.[15] The phosphor must have the capacity to be used under these severe conditions.

Research into phosphor for FEDs is progressing. FEDs are driven one line at a time. Since the distance from the cathode to the phosphor screen is much less than that for CRTs, the acceleration voltage is only several KV. As a result, the excitation current density ranges from one-tenth up to about the same as that for CRTs. As an example, an FED which uses P-22, has a light-emission duty ratio of 1/256, an accelerating voltage of 4 KV and a current density of 15 A/m^2 and yields 600 cd/m^2.[16] The light-emitting efficiency is 8 lm/W.

We now compare an FEL with an FED. Since the distances between the phosphor screen and the cathode are similar, the acceleration voltages are comparable. Considering that light utilization with LCDs is less than 10%, the brightness and emission efficiency of the FEL for backlight applications must be 10-times that of the FED. On the other hand, the entire screen is illuminated simultaneously, making the required current density less than a tenth of that of the FED. Achieving 10 times higher light emission efficiency with FED phosphors compared to FEL phosphors is not necessarily an impractical goal. Generally, the efficiency of cathodoluminescent phosphor increases as the acceleration voltage rises, and as the driving current density decreases.[17]

18.5 Issues Relating to Practical Field Emission Backlights

Difficulties that must be overcome in order to produce practical field emission backlight units are summarized in this section. A key requirement of the field emission source is uniformity. Minute, fibrous, carbon materials such as CNTs are excellent in terms of ease of manufacturing and lifetime, but it is difficult to achieve uniformity. Research on FEDs and similar devices is directed towards this issue.[18] As discussed earlier regarding cathodoluminescent phosphors, a design is needed for obtaining a current density for FELs which is much lower than that for FEDs. The threshold voltage of the field emission with CNTs is somewhat unstable, so that driving the FED emitter with low voltages (i.e. low current density) results in a loss of uniformity.[19] Further development of electron emitter materials for field emission backlights is needed.

For the vacuum container, a method for applying as high a voltage as possible is needed. Also, spacers are required for large-screen backlights. Therefore measures are needed to overcome problems arising from the use of the spacers, including uneven brightness, charge-up of spacer surfaces and discharge breakdown.[7,20] It is not unreasonable to say that the development of efficient cathodoluminescent phosphors is the key to the success of the field emission backlight technology. There have been reports on possible materials,[21,22] and future developments are attracting much attention.

References

[1] Forbes, R. G. (2001) 'Low-Macroscopic-Field Electron Emission from Carbon Films and Other Electrically Nanostructured Heterogeneous Materials: Hypotheses about Emission Mechanism', *Solid-State Electronics*, **45**, pp. 779–808.

[2] Brodie, I. and Richard, P. (1994) 'Vacuum Microelectronic Devices', *Proceedings of the IEEE*, **82**, pp. 1006–1034.

[3] Bonard, J.-M. and Schwoebel, P. R. (2002) 'Field Emission Properties of Carbon Nanohorn Films', *J. Appl. Phys.*, **91**, pp. 10107–10109.

[4] Teo, K. B. K. *el al.* (2004) 'Catalytic Synthesis of Carbon Nanotubes and Nanofibers', *Encyclopedia of Nanoscience and Nanotechnology*, pp. 665–686.

[5] Choi, W. B. *et al.* (1999) 'A 4.5-in. Fully Sealed Carbon Nanotube-Based Field-Emission Flat-Panel Display', *SID '99 Digest*, pp. 1134–1137.

[6] Lee, Y. D. *et al.* (2004) 'Flat Lamp Fabrication Using Double-Walled Carbon Nanotubes Synthesized by Thermal CVD', *Proc. IDW '04*, pp. 1183–1186.

[7] Lee, H. J. *et al.* (2005) 'Field Emission Properties of 4.5 inch Triode Type Flat Lamp using the Screen Printing Method', *SID '05 Digest*, pp. 422–425.

[8] Lee, D. J. *et al.* (2001) 'Flat Lamp Fabrication Using Thermal Grown Carbon Nano Tubes', *Proc. Asia Display/IDW '01*, pp. 1249–1252.

[9] Sugawara, T. *et al.* (2006) 'A Novel Spacer-Free Panel Structure and Glass for FED', *SID '06 Digest*, pp. 1752–1755.

[10] Moon, S. Y. *et al.* (1999) 'Structural Analysis of Vacuum Panel for FED', *Proc. IDW '99*, pp. 927–930.

[11] Curtin, C. J. *et al.* (2000) 'Scaling of FED Display Technology to Large Area Displays', *SID '00 Digest*, pp. 1263–1265.

[12] Tanaka, M. *et al.* (2004) 'Development of a 11.3-inch VGA Full-Color FED', *SID '04 Digest*, pp. 833–835.

[13] Ushirozawa, M. *et al.* (2003) 'Feasibility Study on Graphite Nanofiber FED', *Proc. IDW '03*, pp. 1243–1246.

[14] Phosphor Research Society of Japan (1999) *Phosphor Handbook*. CRC Press, pp. 471–497.

[15] Yamamoto, A. (1997) 'Technical Demands and Trends of FED Phosphors', *Monthly Display*, **3**, pp. 63–67 (in Japanese).

[16] Kusunoki, T. *et al.* (1999) 'Photolithographic Fabrication of an MIM Cathode Array and Color Display Operation', *Technical Report of IEICE*, **EID98-92**, pp. 71–76 (in Japanese).

[17] Kykta, M. (1999) 'Phosphor Requirements for FEDs and CRTs', *Information Display*, **11**, pp. 24–27.

[18] Dijion, J. *et al.* (2006) 'A Status on the Emission Uniformity of CNT FED Technology', *SID '06 Digest*, pp. 1744–1747.

[19] Ushirozawa, M. *et al.* (2004) 'Investigation of Graphite Nanofiber as an Emitter for FED', *Proc. IDW '04*, pp. 1218–1220.

[20] Tirard-Gatel, N. *et al.* (1999) 'Charging and Reliability Effect Associated with FED Spacers', *SID '99 Digest*, pp. 1138–1141.

[21] Takigawa, Y. *et al.* (2005) Fabrication and luminescent properties of white phosphors for field emission lamps (I)', *Extended Abstracts (The 52nd Spring Meeting, 2005); The Japan Society of Applied Physics and Related Societies*, p. 1624 in Japanese).

[22] Chi, E. J. *et al.* (2006) 'Recent Improvements in Brightness and Color Gamut of Carbon Nanotube Field Emission Display', *SID '06 Digest*, pp. 1841–1844.

Part Three
Optical Components

19

Light-guide Plates

Y. Ishiwatari

Asahi Kasei Chemicals

19.1 Introduction

Manufacturing of polymethylmethacrylate resin (hereafter denoted as PMMA) in Japan began in 1938. Since then Japan has had a long history of plastic industries for over half a century. In the beginning, plastics were used only for windshields for airplane cockpits. Since World War II, a wide variety of utilization has been developed with the rapid growth of the economy. In particular, PMMA has many incomparable characteristics such as high optical transmission, glossiness and excellent endurance to weather and radiation. Thus PMMA is widely used both indoors and out of doors.

PMMA is supplied as a plate type or a mold type. The plate type is processed by mechanical processes (cutting, polishing), heat process (molding), adhesion process and printing. With these processes, PMMA is used for a wide variety of applications such as digital signs, covers for lighting, coin venders and exteriors of automobiles. Recently the light-guide plates and light-diffusing plates in LCD-TV backlights – as well as in large-area screens for projection displays – also started using PMMA. Thus the volume of utilization of PMMA is rapidly growing. PMMA is also used as pellets for injection molding and extrusion molding. These are used to manufacture the tail lamps of automobiles, mechanical parts, vessels, bottles, lenses and

LCD Backlights   Edited by Shunsuke Kobayashi, Shigeo Mikoshiba and Sungkyoo Lim
© 2009 John Wiley & Sons, Ltd.

pipes. This chapter is devoted to the market demands, fundamental optical characteristics and required properties of PMMA.

19.2 Market Demands for PMMA

19.2.1 PMMA Plate

Light-guide plates for notebook PCs fabricated by injection molding have been widely adopted due to the advantage of their low weight. For LCD PC monitors with a 17-inch diagonal or larger, properly shaped PMMA plates are used in order to obtain high luminance. LCD monitors are thinner than CRT monitors, and LCD worldwide volume sales overtook those of CRTs in 2004. The desk top and notebook volume was 290 million sets altogether in 2007 with an annual growth rate of 25%. The worldwide volume sales of LCD-TVs was 93 million sets in 2007 with an annual growth rate of 39%.

Along with the growth of these products, manufacturers planned to produce more PMMA plates. Asahi Kasei Chemicals inaugurated a factory operating with an annual production of 6000 tons in Korea in 2003. Kuraray produced 5000 tons of injection molded plates in 2004, and they also started producing 3000 tons of cast plates annually in China in 2005. Sumitomo Chemicals set up their domestic production facility to 5000 tons annually, and also another 5000 tons annually in China in 2004. In 2005, Mitsubishi Rayon started the commercial production of 20 000-ton per annum of continuous casting plates annually in China.

19.2.2 Injection Molding Materials

In 2004, the demand for PMMA for automobiles, home appliances and IT products reached 127 000 tons in accordance with a steady growth of the light-guide plates for LCDs. In IT-related products, the rapid growth of light-guide plates should be noted. The total volume of note book PCs, PC monitors and mobile terminals in 2007 was over 290 million units with an annual growth of 25%. In particular the market for LC-TVs and notebook PCs was growing steadily. This trend brought an increase in the demand for injection molding materials.

With regards to the region of production, the center has shifted from Japan to Korea, Taiwan and China. Along with the growth in the demand, Japanese chemical companies were establishing factories for raw materials as follows. Mitsubishi Rayon started a new factory of 40 000 tons annually in 2003 in China. Kuraray increased domestic annual production to 20 000 tons in 2004.

Sumitomo Chemicals also increased annual production to 45000 tons in Singapore in 2005, and plans to produce up to a further 50000 tons annually in Singapore. Asahi Kasei Chemicals also increased the domestic annual production by 5000 tons in 2004.

19.3 Characteristics of PMMA

PMMA has superior optical properties compared with PC (polycarbonate), PS (polystylene), COP (cyclophreon) and other plastic materials as shown in Table 19.1. PMMA has the excellent characteristics of high optical transparency, high durability for weather conditions and ambient lights and low refractive index.[1–4]

19.3.1 Optical Transparency

PMMA has an optical transmission of 93% in the visible range. This superior value originates from its low refractive index which results in low surface reflection. For this characteristic, PMMA has the highest transparency of all plastics.[5] PMMA is a polymer having the molecular structure shown in Figure 19.1. PMMA is synthesized by polymerization of MMA monomers using a radical catalyst. Its transparency is the most important optical characteristic, affecting reflection, absorption and scattering.

19.3.1.1 Optical Reflection

The optical reflection coefficient, R, is described by the Fresnel equation,

$$R = (n-1)^2/(n+1)^2,$$

Table 19.1 Comparison of properties of transparent resins with PMMA.

	PMMA	PC	PS	COP
Total optical transmission (%)	93	87	89	90
Refractive index (-)	1.49	1.59	1.59	1.53
Saturation water absorption ratio (%)	2.0	0.2	0.05	0.01
Water vapor transmission (g/m²·24h)	41	44	30	15
Weight bending temperature (°C)	74–103	130–141	65–104	123

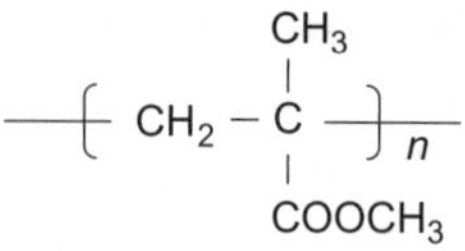

$$CH_3$$
$$-\!\!\!-\!\!\left[\!-\ CH_2 - C\ -\!\right]_n\!\!\!-\!\!\!-$$
$$COOCH_3$$

Figure 19.1 Chemical structure of PMMA.

Table 19.2 Optical absorption by electron transition.

Resins	Optical absorption by electron transition (dB/km)		
	600 nm	700 nm	800 nm
PMMA	0.0	0.0	0.0
PC	250	110	60
PS	210	60	20

where n is the refractive index. R is reduced as n becomes smaller. The reflection coefficient of PMMA is 7% with $n = 1.492$. The optical loss of PMMA mostly occurs from surface reflections and the internal loss is very small.

19.3.1.2 Optical Absorption

When light passes through a medium, the constituent molecules absorb the light, which induces rotation of the molecules, interatomic resonant vibrations and the movement of electrons. The transparency in the visible range is governed by the binding strength between electrons, where the single bond (based on σ electrons) is the strongest and the double bond (through π electrons) is weaker. PMMA contains mainly single bonds, giving the highest transparency as shown in Table 19.2. Figure 19.2 demonstrates the reduction of transmission of PMMA by photon absorption occurring in the short wavelength region below 340 nm. Transmission decreases with increasing thickness of a plate.

19.3.1.3 Optical Scattering

Another loss comes from the optical scattering that is inherent to polymer materials. Scattering is also caused by the inclusion of impurities. The fundamental scattering phenomenon is Rayleigh scattering which is expressed as

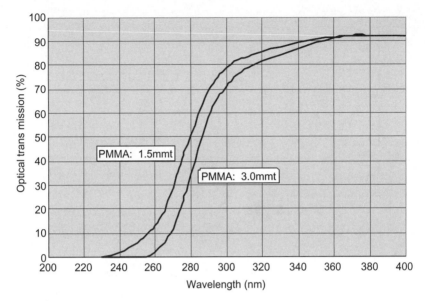

Figure 19.2 Optical transmission spectra of PMMA in the UV range.

$$\alpha = 2\pi^3 \left(\Delta n/n\right)^2 V \, v/\lambda^3,$$

where α, Δn, V, v and λ are the scattering loss coefficient, the difference in refractive indices of the scattering particles and the host medium, the volume of the scattering center, the volume occupation ratio of the scattering centers and the wavelength. Strong scattering occurs as Δn increases and λ decreases.

19.3.2 Durability in the Environment and UV Light

PMMA has excellent durability against UV exposure, since the α-position of MMA, which is the main constituent, is substituted with a methyl moiety. If necessary the durability can be further improved by doping the PMMA with UV absorbing agents.

19.3.3 Refractive Index

PMMA has a low refractive index and is optically isotropic, since it does not contain highly polarizing constituents such as benzene rings, which are found in PC and PS.

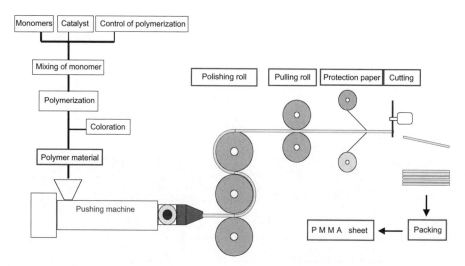

Figure 19.3 Injection molding method.

19.4 Manufacturing Method for PMMA Plates

PMMA plates are manufactured by either injection molding or cast molding. Injection molding is used mostly for its high productivity with no need for a large investment. Cast molding is divided into glass-cell cast and continuous cast methods. These methods have the advantage that PMMA has high solubility in solvents, enabling the use of solvents which have large molecular weights. The methods, however, have the disadvantage of low productivity. The continuous cast method has an advantage compared with the glass-cell method in the relative simplicity of mass production, but it needs a large-scale production facility. Figures 19.3 and 19.4 illustrate injection molding and continuous cast molding, respectively.

19.5 Applications to LCD Backlight Units

As mentioned above, PMMA is a plastic that has low reflection, high transmittance due to low Rayleigh scattering loss and low wavelength dependency of refractive index owing to its chemical structure. PMMA is used more and more in a variety of optical materials, in particular in LCD backlight units. The light-guide plates for LCD backlight units will be explained below.

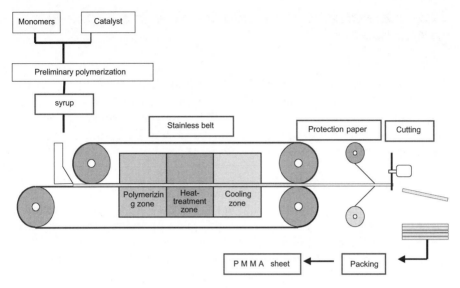

Figure 19.4 Continuous casting method.

19.5.1 Types of Backlight Units

LCDs need backlight units, since LCDs are not emissive displays. There are two types of backlights. One uses lamps behind an LCD panel. Between the lamps and an LCD panel a light diffusion plate, a diffusion film, a lens film, etc., are installed. It is necessary for the unit to have a gap between the lamp and the diffusion sheet, otherwise images of the lamps may appear on the LCD screen. This limits the ability of the backlight unit to be made as thin as possible.

The other type of backlight is an edge light unit in which a CCFL lamp is installed at the edge of a light-guide plate. This unit has the advantage of being thin. However, it becomes difficult to have uniformity and high luminance as the area of the backlight unit increases. In order to overcome this, a thick and heavyweight light-guide plate is used, but this makes it difficult for wide spread use of the edge light unit in large displays. In order to attain an optimum trade-off between high luminance and low power consumption, intensive research and development into lamps, optical films and lightguide plates are underway.

19.5.2 Operating Principle and Structure of an Edge-light Type of Backlight Unit

19.5.2.1 Lamps

There are two kinds of lamp: cold cathode fluorescent lamps (CCFLs) and hot cathode fluorescent lamps (HCFLs). Of these, only CCFLs are used for edge light backlights due to their long life, low power consumption, low heat dissipation and thin tubes. In parallel with this, backlights using LEDs are being actively investigated.

19.5.2.2 Light-guide Plates

The operating principle of the edge-light type backlight unit is as follows. Light emitted from a CCFL enters a light-guide plate from the front edge. The light is propagated along the light guide by total internal reflection. When the light hits a dot pattern formed on the back side of the light-guide plate, part of the light emerges from the light-guide plate provided that the incident angle is smaller than the critical angle of 42.2°. The dot pattern has to be designed so as to give a uniform illumination from the plate.

Figures 19.5 and 19.6 illustrate the operating principle of the edge-light unit and an example of the unit structure, respectively. Figure 19.7 shows a photo of a 17-inch monitor. There are several types of light-guide plates: (1) a dot pattern is formed with white ink on one side of the cast mold plate; (2) a textured pattern is formed using an injection molding process; (3) an edge pattern is formed using an injection molding process.

19.5.2.3 Diffusion Film

A diffusion film looks like frosted glass. It is fabricated by dispersing light-scattering particles into a film or by coating diffusion particles onto the film.

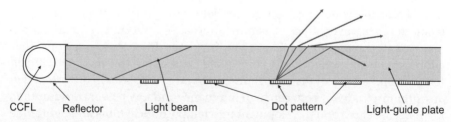

CCFL   Reflector   Light beam   Dot pattern   Light-guide plate

Figure 19.5 Edge-light guide: refractive index of air, $n_1 = 1.00$; refractive index of acrylate, $n_2 = 1.49$; critical angle $\theta_c = \sin^{-1}(n_1/n_2) = 42.2°$.

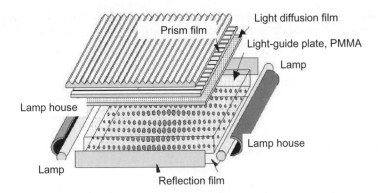

Figure 19.6   Example of edge-light type backlight unit.

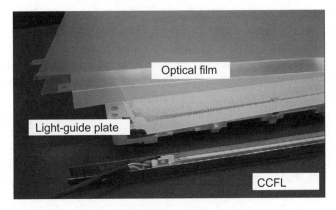

Figure 19.7   Structure of a 17-inch monitor.

The diffusion films play a role in eliminating the dot patterns from appearing on the LCD display.

19.5.2.4 Reflection Film

White color PET films are widely employed. The utilization of light is enhanced by stacking the reflection films on the back side of the light-guide plate, on which the dot pattern is printed.

19.5.2.5 Prism Films

Without the prism films, light from the backlight unit propagates not only in the perpendicular direction but also at oblique angles. By using prism films the perpendicular component can be increased by 20%. For the prism films, PET films having a fine prism pattern are commonly used.

19.6 Characteristics Required for Materials of Light-guide Plates

There are two kinds of PMMA materials: one is a plate type, and the other is a pellet type used for injection molding.

19.6.1 Requirements for Plate and Pellet Mold Materials

19.6.1.1 No Extra Inclusions

Inclusions in the light-guide plates are the cause of mura (non-uniformity of luminance), bright spots and dark spots. When pellets are used, attention should be paid to inclusions coming from burned waste in the cylinder, extra resins, extra trash arising during the process of charging the material and burned wastes which adhere to the sliding part of the metal cast.

19.6.1.2 Colorless

Colorization of resins should be avoided since this results in color mura. For pellet molding, the plate becomes yellowish when there are residues of resin and depleted wastes in the cylinder. Sometimes dusts ingress during the molding process become the cause of an extra color formation. Invisible dust, which is accumulated in a hopper dryer, may cause degradation of the characteristics.

19.6.1.3 Excellence in the Optical Transmission over a Long Distance

PMMA is an excellent optical transmitting material. Nevertheless it is necessary to check the optical loss for light traveling a long distance, which may be degraded by colorization.

19.6.1.4 Durability at High Temperatures

The light-guide materials must withstand a temperature of 80 °C, since the light-guide plates are positioned close to lamps to reduce the thickness of the backlight unit.

19.6.2 Characteristics Required for Plate Materials

PMMA plates are supplied in the form of sheets. After they are cut, the dot pattern is printed on the surface. When the plates are being printed, uniformity of the patterns is required. Non-uniformity often arises from a curvature or uneven thickness of the plates. It is necessary to provide a flatness accuracy of ±0.1 mm. Generally, PMMA has the property of absorbing water. Even a small amount of water absorption may cause a change in the dimension of a PMMA plate. Therefore it is necessary to cover the products by wrapping them with a high-quality film.

19.6.3 Required Characteristics for Pellets

19.6.3.1 High Fluidity

When fabricating a thin light-guide plate with a large area and having a prescribed shape, sufficient pressure should be uniformly applied to the material. Thus using a material having a high fluidity is desirable. The use of a high fluidity material also makes the plate free from deformation and cracks arising from printing inks, since the internal strain becomes less.

19.6.3.2 Low-temperature Casting

Casting at a low temperature yields desirable properties. As the molding temperature is lowered, the strain at the surface of the product becomes smaller, reducing bending of the plate. Also low-temperature production reduces coloration and degradation due to heating. However, a low-temperature process provides low fluidity. During the process of molding, the outer parts are cooled down first. Due to the low fluidity, only the outer parts become solidified. Then next, the inner parts are cooled down. This causes volume shrinkage in the inner part producing a volume density difference; this results in a higher refractive index in the outer part so that the light beams tend to be guided to the outer part. If we look at a light-guide plate from the opposite end, the image of an object looks smaller. Considering these effects, the low-temperature casting process should be controlled so as to assure the luminance balance.

19.6.3.3 Ease of Removing a Molded Material from a Cast

If the removal of a molded product from a cast is difficult, then the product yield becomes low due to deformation of the surface.

19.6.3.4 Protection from Humidity

Damp pellets produce silver-colored defects during the molding process. For this reason, damp pellets should be dried thoroughly in advance.

19.7 Materials for Extrusion Molding and Injection Molding

19.7.1 Materials for Extrusion Molding

The requirement for a light-guide plate is to guide the light through the plate and then to illuminate the whole the area of the plate uniformly without a significant loss of light. PMMA satisfies these requirements. Products, however, must be manufactured with a high level of quality control to eliminate any reduction in the luminance level and the appearance of bright spots which originate from the inclusion of dust or particles. Reduction in the internal strain is also important since it causes bending failure during environmental testing.

In order to provide a uniform optical output, a small amount of fine particles is doped uniformly in a light-guide medium as shown in Figure 19.8. The fine particles cause an enhancement of the optical throughput in the upward direction. Figure 19.9 demonstrates the result obtained by using light-scatting particles. Recently, research and development of advanced backlight units have been directed towards the integration of a light-guide plate with high-grade films such as light-diffusion films, converging lenses and reflection films. Advanced technologies such as multilayering, shaping and optical scattering have also been investigated.

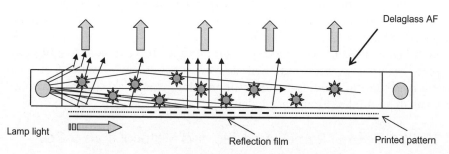

Figure 19.8 Scattering-type light-guide plate.

Light-scattering type   **Conventional type**

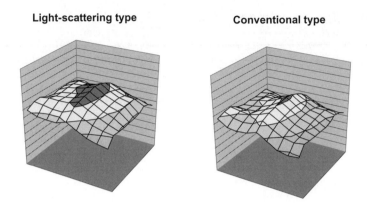

Figure 19.9   Profile of luminance levels: comparison between a light-scattering type and conventional type.

19.7.2 Materials for Injection Molding

The basic physical properties of materials used for injection molding are characterized and standardized by 70 NH and 80 NH, as shown in Table 19.3. Materials which satisfy these standards must be free from degradation and colorization caused by UV from CCFLs. Also these materials should have a good balance of fluidity and heat resistance while the product is being removed from the metal cast.

19.8 Conclusions

PMMA has excellent properties such as high optical transparency, durability under extreme weather conditions and light attack, and a beautiful surface glossiness. For these reasons PMMA will be used more and more in various application fields of optical technologies such as information display, optical recording, optical communication, diffusion film for LCD-TVs, DVD substrates, plates for PTV and PDP, LED lamps for automobiles and general lighting in the home. PMMA will be converted into monomers almost 100% by heating, meaning that PMMA can be completely recycled. The demand for PMMA will increase because of these characteristics.

Table 19.3 Fundamental physical properties of light-guide plate of injection molding grade.

Physical properties	Testing method	Units	Heat proof high fluid grade 70NH	Heat proof good fluid grade 80NH
			Heat proof, high fluidity Light proof, molding	High heat proof good fluidity light proof molding
1. Rheological properties				
Fluidity/MFR	ISO1133 (cond.13)	g/10 min	10.5	5.5
2. Mechanical properties				
Tensile Modulus	ISO 527-2/ 1A/1	MPa	3200	3200
Tensile Strength	ISO 527-2/ 1A/5	MPa	67	72
Tensile Strain	ISO 527-2/ 1A/5	%	5	5
Charpy impact strength No notch Strength	ISO 179/1eU	KJ/m^2	19	20
Charpy impact strength With notch Strength	ISO 179/1eA	KJ/m^2	1.2	1.3
3. Heat proof properties				
HDT	ISO 75-1 75-2	°C	97	100
Vicat softening point	ISO 306 B 50	°C	105	109

References

[1] Takeda, K. (1979) *Plastic Material Technology*. Sigma Publishing (in Japanese).
[2] Ide, F. (1995) *Optoelectronics and Polymer Materials*. Kyoritsu (in Japanese).
[3] (1988)*Advanced Polymer Materials, Technical Overview*. Bikohsha (in Japanese).
[4] (1986)*Polymers Data and Handbook*. Place: Baifukan (in Japanese).
[5] Koike, Y. (1994) *Optical Properties of Polymers*. Place: Japanese Society of Polymers (in Japanese).

20

Optical Diffuser Plates

Y. Ishiwatari

Asahi Kasei Chemicals

20.1 Introduction

Liquid crystal displays are divided into the following categories: mobile phone, PDA, car navigator, notebook PC, desktop PC and LC-TV. For the categories from the mobile phone to the desktop PC, the use of diffuser plates is mandatory to convert point light sources (LEDs) or line light sources (CCFLs) into a flat light source. Materials used for the diffuser plates are PMMA (polymethylmethacrylate), MS (styrenemethacrylate), MMA (methacrylacidmethyl), PS (polystyrene), PC (polycarbonate) and COP (cyclophreon), but PMMA ranks as the de facto standard because of its excellent optical quality. While the CCFL is widely used as a light source, LEDs, HCFLs, EEFLs and FFLs are also being developed and some of them are already commercialized.

Figure 20.1 illustrates the structure of a backlight unit for LCD-TVs. Table 20.1 lists the major functions and properties of the materials required for the diffuser plate. The plate is indispensable for a flat light source, diffusing the light which comes from CCFLs located beneath the plate. The plate also cuts UV radiation and heat, preventing yellowing of the plate and over-layered films. The plate has to have rigidity to sustain the weight of peripheral optical films, and therefore the plate must have an appropriate thickness. When the diffuser plate is exposed to humid conditions during transportation or storage,

LCD Backlights   Edited by Shunsuke Kobayashi, Shigeo Mikoshiba and Sungkyoo Lim
© 2009 John Wiley & Sons, Ltd.

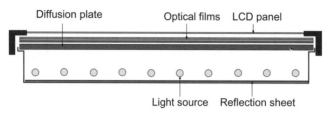

Figure 20.1 Structure of a backlight unit.

Table 20.1 Functions and properties of materials required for diffuser plates. O: good, △: acceptable, ×: poor. PMMA: polymethylmethacrylate, MS: styrenemethacrylate, PS: polystyrene, PC: polycarbonate.

	PMMA	MS	PS	PC	tasks
optical characteristics	O	△	△	△	luminance
color	O	△	△	×	colorization
change of dimension by water absorption	×	△	O	O	warp
surface hardness	O	O	O	×	scratch
rigidity	O	O	O	△	flexure

water vapor absorbs into the plate up to its saturation level. If CCFLs are turned on under such saturated conditions, the CCFL side of the plate dries and shrinks, resulting in concave bending of the plate. The plate may then push against the LCD panel, causing mura (non-uniformity) on the screen.

The diffuser plate is separated from the CCFLs by supporting pins. The pins are made of resin doped with inorganic pigment, and they have sharp tips to eliminate formation of their shadows on the LCD screen. Since the tips of the pins are in direct contact with the diffuser plate, this may cause scratches if vibrations occur during transportation.

20.2 PMMA Light Diffuser Plates

Diffuser plates made of PMMA have the advantages of excellent optical transparency, durability against radiation, surface hardness (scratch free), rigidity and strength. Furthermore, the plates can be installed close to light sources, since PMMA has only a slight yellowing tendency when exposed to UV irradiation. Inorganic or organic light-scattering agents (particles) are sometimes doped and dispersed in the host resin of the diffuser plate to control the total transmission and light scattering. Figure 20.2 shows the

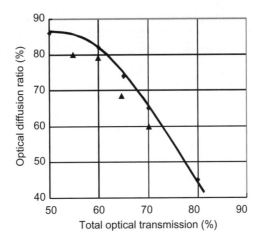

Figure 20.2 Relationship between the optical diffusion ratio and the total optical transmission.

relationship between the optical diffusion ratio (%) and the total optical transmission (%). The horizontal axis represents the light output and the screen luminance becomes higher with increased transmission. When the optical diffusion ratio increases an image of the light sources becomes less visible on the LCD screen. It is therefore necessary to have a compromise between transmission and diffusion.

Another important design parameter is the size of the diffuser plate. This becomes especially important for large-size backlight units. Degradation of luminance uniformity caused by the expansion or shrinkage of the plate due to changes in temperature, as well as bending due to humidity, should be minimized. Although PMMA has excellent optical properties, it has a relatively high tendency for bending when it is subjected to humidity. One of the methods to overcome this is to coat the surface with a low water-absorbing layer. For LCDs with diagonals of more than 30-inch, demand for the use of PMMA is declining for this reason.

20.3 MS and PS Light Diffuser Plates

MS resin is a copolymeric compound of MMA and ST (styrene) which has low water absorbent properties. There are a few resins consisting of copolymeric compounds. Generally, a resin with a higher styrene content has lower water absorption. MS resin has both the excellent optical properties of

PMMA and the low absorbent properties of PS. Therefore, the demand for MS resin is increasing along with the growth of large-size LC-TVs. Further, the demand for PS is also increasing because of its low bending characteristics with water absorption and its high rigidity. Figure 20.3 compares the amount of warp for both PMMA and PS with respect to testing time. Figure 20.4 explains the mechanism of the bending.

PS resin is used for a wide variety of goods due to its cost effectiveness. Although PS is generally inferior to PMMA and MS resins in terms of durability against irradiation and mechanical shock, commercially available PS diffuser plates have eliminated these defects. When bending of the plate due to absorption of water is highly undesirable, COP or PC resins are used. COP

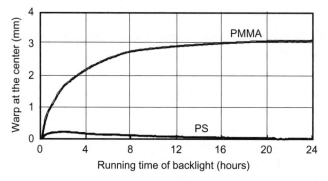

Figure 20.3 Warp of PMMA and PS measured at the central position of the plates with respect to the running time of the backlight.

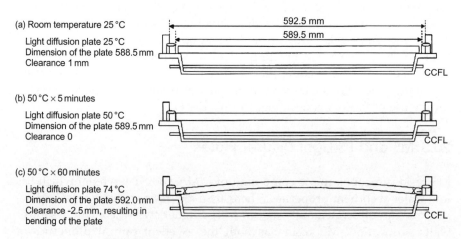

Figure 20.4 Mechanism of bending.

is a ring-shaped polymer using cycloolefins as monomers, and it has high optical transparency and low absorption of water due to the nature of the olefins, as well as high durability to the environment. COP is used for optical components such as lenses and fibers. Taking advantage of the low absorbent nature of COP, it is also used as an injection-molding material for medium sized LC-TVs. Recently the adoption of COP for diffuser plates for large sized LC-TVs has been discussed. The method used an injection mold or special metal casting by improving its fluidity. However, fabrication of thin plates, improvement of rigidity, reduction in costs and stability of mass production should be discussed further.

PC is an aromatic compound and it is an engineering resin which features a high optical quality owing to low absorbtion and high temperature resistance. PC, however, is inferior to PMMA, MS and PS as far as mechanical rigidity is concerned, causing bending to occur in thin diffuser plates. PC should be improved further to provide radiation durability and strength against scratching.

20.4 Trends in Light Diffuser Plates

Although there are various materials which can be used for light diffuser plates, each material has its own merits and demerits, and this may be attributed to different requirements by different LC-TV manufactures. This also may be due to the recent rapid growth of LCD-TVs and to the rapid decline of their price. Recently value-added products have been demonstrated in which an optical component such as a prism sheet was integrated into the diffuser plate. Figure 20.5 shows the concept of integration for a diffuser plate with other optical films such as a prism sheet.

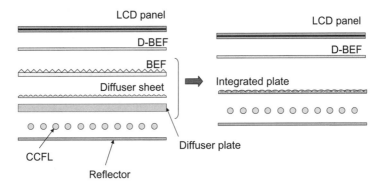

Figure 20.5 Concept of an integrated plate.

20.5 LED Light Sources and Diffusing Plates

LEDs can be used as alternatives to CCFLs. LEDs have a wide color gamut, long life and high luminance, but it may take more time for them to be used extensively in LCD-TVs because of their high cost. LEDs have already been introduced into mobile LCDs. The same light diffuser plates are used for LEDs and CCFLs. In the future, a more appropriate diffuser system for LED light sources is likely to be developed.

21

Lens Films and Reflective Polarization Films

F. Hanzawa

Sumitomo 3M

21.1 Introduction

The technologies of lens films and reflective polarization films will be explained in this chapter. These films enhance the brightness and reduce the power consumption of backlight units for direct-view LCDs.[1,2]

21.2 Fundamentals of Reflection and Refraction

Light traveling from a medium with a high refractive index to a low refractive index experiences total reflection when the incident angle exceeds the critical angle. The case of 100% reflection is called total internal reflection (TIR), in which case the light is confined within the medium having the higher refractive index. This phenomenon is used in optical transmission fibers and light-guide plates in LCD backlight units.

Figure 21.1 shows reflectance as a function of the incidence angle when light travels from acrylate (with $n = 1.49$) into air ($n = 1.0$). Here, R_p and R_s

LCD Backlights   Edited by Shunsuke Kobayashi, Shigeo Mikoshiba and Sungkyoo Lim
© 2009 John Wiley & Sons, Ltd.

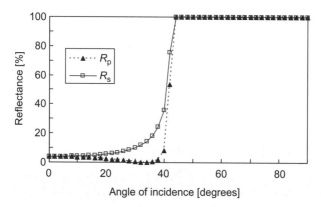

Figure 21.1 Reflectance of p-polarized and s-polarized light, R_p and R_s, travelling from acrylate (n = 1.49) into air (n = 1.0).

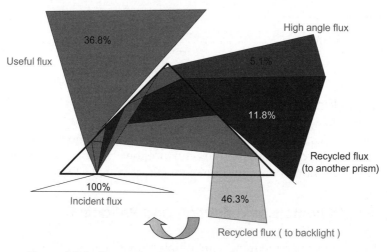

Figure 21.2 Recycling of wasted light by the use of a BEF.

are, respectively, the refractive indices of the p-polarized light (in the plane) and the s-polarized light (at right angles to the p-polarization). The lens films are designed based on Fresnel's law and fabricated using a resin with high transparency and high refractive index. Percentages of interfacial reflection and refraction are illustrated in Figure 21.2. The useless component of the flux is recycled and changed to 'useful flux' in the backlight/prism and returns to the LCD, enhancing its brightness.

21.3 Lens Films (Upward Direction)

21.3.1 Lens Films (BEFs)

As one of the typical examples of a lens film, the brightness enhancement film (BEF) of Sumitomo 3M will be described. A prism pattern is formed with optical precision on a polyester film having a high optical transparency. The film is attached to the front surface of a backlight unit and is capable of enhancing the brightness by recycling light which is normally wasted. The recycling is achieved by the above-mentioned relations with regard to the difference in refractive index and incidence angle. Figure 21.3 shows the luminance gain with respect to the observation angle. In the diagram, 'BEF II 90/50(V)/BEF II 90/50(H)' denotes that two orthogonally arranged films are stacked, resulting in a luminance gain of 2.2. The lens films play a role in converging the light into the normal direction to the plane of the backlight unit. Usually the BEF is placed on top of a stack of films which consists of a light-guide plate, a reflection sheet and a diffusion sheet. The surface of the lens is facing towards the LCD (upward direction). There is also the case in which the lens surface is facing towards the backlight unit (downward direction) as will be mentioned later.

21.3.2 Lens Films with Round-tipped Lenses (RBEF)

The lens shown in Figure 21.2 has a sharp tip. The round brightness enhancement film (RBEF), on the other hand, has a lens whose tips are round which makes the variation of the luminance gain with respect to the observation

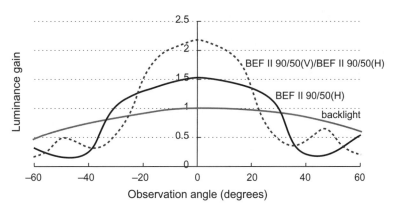

Figure 21.3 Luminance gain versus the observation angle for a BEF II.

angle less steep. Thus the RBEF gives a compromise between high luminance gain and a wide viewing angle.

21.3.3 Random Pattern Lens Films

In order to eliminate optical interference between the lens system and the pixel arrangement of an LCD, BEFs are fabricated whose lenses are arrayed in a non-linear fashion.

21.3.4 Waved Films

The waved films increase the viewing angle while enhancing the luminance. The cross section of the film has a wavy profile instead of a sharp edge. The film is a single polycarbonate plate which is scratch free, so that a protecting film is not required.

21.3.5 Upward Direction Lens Films

Various lens films which enhance brightness and increase the viewing angles are summarized in Figures 21.4 and 21.5. Also nonuniformity of the light

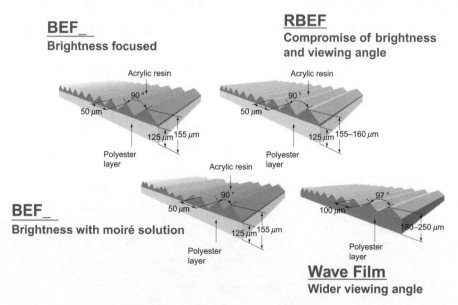

Figure 21.4 Various optical lens films.

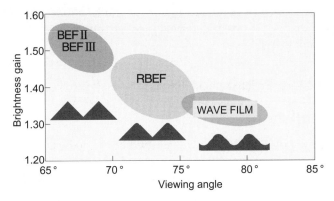

Figure 21.5 Luminance gain versus the viewing angle.

| Diffuser plate | Diffuser plate +diffuser sheet | Diffuser plate +diffuser sheet +BEF |

Figure 21.6 Images of CCFL lamps appearing on an LCD screen.

output is eliminated by recycling the light proceeding towards the backlight unit. Figure 21.6 demonstrates how the images of the fluorescent lamps appearing on the screen are reduced by using the lens films.

21.4 Lens Films (Downward Direction)

Lens films, whose lens is facing downward, have the capability of enhancing the brightness in the front direction, but with a reduction in viewing angle. The films are mostly used in the side-light type. A careful design is needed for an adequate combination of lens films and the light-guide plate, since the dependence of the luminance on the direction is strong. Recently, an advanced BEF having asymmetric lenses was proposed[3] for which an even stronger directivity was obtained. In some mobile phones and PDAs with an LED light source, prisms are arrayed in an arc determined by the distance from the light source.[4]

21.5 Reflective Polarization Films

21.5.1 Functions

Although the p-polarized light is transmitted through the LCD rear polarizer into the LCD, the s-polarized light is absorbed. The loss can be recovered by inserting a film between the backlight unit and the rear polarizer and converting the s-polarization into p-polarization. The principle of recycling the light is illustrated in Figure 21.7. When a DBEF is not used (normal BL), the entrance polarizer transmits the p-polarized light but s-polarized light is absorbed. On the other hand, with a DBEF, s-polarized light is reflected back to the backlight unit without being absorbed and depolarized by the diffusion and scattering. The new p-polarized component is transmitted to the LCD but the s-polarized light is reflected again to the backlight. The film has the advantage that no reduction of the viewing angle occurs.

A dual brightness enhancement film (DBEF) exhibits a luminance enhancement of 60%. By combining it with a lens film, further enhancement can be

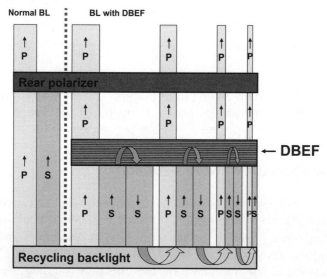

Figure 21.7 Utilization of light with and without a DBEF.

obtained. The BEF and DBEF are applicable to all LCD monitors, car navigation systems, PDAs, digital cameras and so on. Also the luminance enhancement results in a reduction of the power consumption and a lengthening of the battery life for portable instruments.

21.5.2 Structure of a Reflective Polarization Film

Reflective polarization films are formed by stacking resins which are optically anisotropic. The thickness of each layer is controlled to an accuracy of the wavelength of light. The structure of the DBEF and the reflection spectra for x and y directions are shown in Figure 21.8.

21.5.3 Example of Applications

DBEFs tend to become thermally deformed because of the inherent nature of the resins used. To overcome this deficiency, DBEFs are fabricated by laminating them with absorbing polarizers or polycarbonate light diffusion films.

21.6 Resin-type Specular Reflection Films

If the optical anisotropy is removed from the above mentioned reflective polarizer film, then it becomes a mirror film in the visible range. This makes it possible to produce a mirror with only a resin sheet. A dual enhanced specular reflector–metal sheet (DESR–M), a reflector for fluorescent lamps, is made by gluing the resin on a metal film.

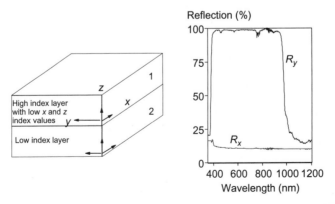

Figure 21.8 Film structure and reflection spectrum of a DBEF.

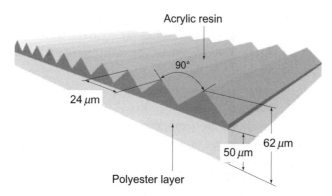

Figure 21.9 Structure of a thin BEF.

21.7 Applications of Films

21.7.1 PDAs

In order to reduce the power consumption, lens films are adopted in most cell phones and PDAs. As shown in Figure 21.9, the base film is thin and the pitch of the prism is small to reduce the total thickness. A matte finish may be used on the rear surface to eliminate the appearance of a moiré pattern arising from optical interference beween the LCD pixels and the prism.

21.7.2 Notebooks

The downward-directed films[4] are frequently used for notebooks because of their luminance and thickness requirements. Upward-directed films, which do not need matching with the light-guide plate, are also employed.

21.7.3 Monitors and TVs

Considering the ease with which large-sized films can be handled, as well as their durability against heat and humidity, relatively thick lens films are used for monitors and TVs. Large-sized TVs often use the RBEF to ensure a wide viewing angle. Figure 21.10 shows thick lens films with 250 μm thick polyester films. For reflective polarization films, polycarbonate diffuser sheets are laminated on DBEF, as shown in Figure 21.11.

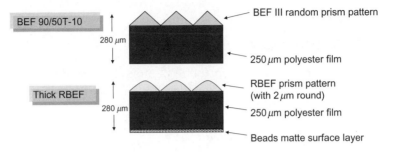

Figure 21.10   Thick lens films for large-sized LC-TVs.

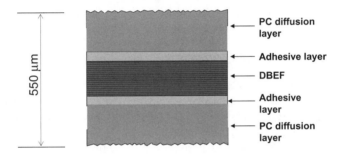

Figure 21.11   Thick reflective polarization films for large-sized LCD-TVs.

Figure 21.12   Structure of a brightness enhancement film-reflective polarizer (BEF–RP).

21.7.4 Combination Films

The use of combination films reduces the number of components and also the production cost. Figure 21.12 illustrates an example of a brightness enhancement film–reflective polarizer (BEF–RP) which is made by combining a thin BEF and a DBEF.[2]

21.8 Standards

21.8.1 TCO '99–03 for FPD Monitors[5]

The TCO '99–03 recommendation for FPD monitors was documented by TCO Development (The Swedish Confederation of Professional Employees). The recommendation limits the vertical and horizontal viewing angles.

21.8.2 TCO '05 for Notebook PCs[5]

Also recommended by TCO Development is a limitation of the half-luminance angle for vertical and horizontal directions. Attention should be paid when using a lens film that reduces the viewing angle by collimating and orienting the light to the front direction.

21.8.3 Environmental Concerns

Environmental concerns have to be considered with the increase in the number of TV sets which are used worldwide. After the Kyoto protocol came into effect on February 16, 2005, CO_2 emissions had to be reduced. Also RoHS, enacted as of July 1, 2006, restricts the use of mercury. Figure 21.13

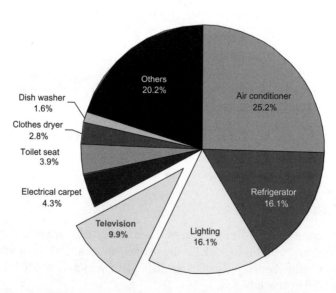

Figure 21.13 A break down of energy consumption of household appliances.

shows percentage of the total energy consumptions of various home appliances. Air conditioners, refrigerators and general lighting are the worst three appliances. Also the energy consumption of LC-TVs is increasing. It is possible to reduce the energy consumption of LC-TVs by adopting a reflective polarization film. This, at the same time, reduces the number of fluorescent lamps used, and hence there is a reduction in the mercury used.

References

[1] Maeda, K. (2003) 'Integration and hybridization of the luminance enhancement films', *FPD International Seminar, E-3*.
[2] Nakahisa, Y. (2003) 'Luminance enhancement films', *The 85th Spring Meeting, Japan Chemical Society* (in Japanese).
[3] Okada, H. (2004) 'Trends in the R & D of lens films for LCD backlights', *Display Monthly*, No. **4**, pp. 14–21.
[4] Makuta, I. (2004) 'Characteristics of backlight, fabrication process, periphery structures', *Gijutsu Joho Kyokai Seminar*.
[5] http://www.tcodevelopment.com/

Index

LCD Backlights   Edited by Shunsuke Kobayashi, Shigeo Mikoshiba and Sungkyoo Lim
© 2009 John Wiley & Sons, Ltd.